Studies in Computational Intelligence 476

For further volumes:
http://www.springer.com/series/7092

Gourab Ghoshal, Julia Poncela-Casasnovas, and
Robert Tolksdorf (Eds.)

Complex Networks IV

Proceedings of the 4th Workshop on Complex Networks CompleNet 2013

Editors
Dr. Gourab Ghoshal
Department of Earth and Planetary Sciences
Harvard University
Massachusetts
USA

Prof. Dr.-Ing. Robert Tolksdorf
Institut für Informatik
Freie Universität Berlin
Berlin
Germany

Dr. Julia Poncela-Casasnovas
Chemical and Biological Engineering
Northwestern University
Illinois
USA

ISSN 1860-949X e-ISSN 1860-9503
ISBN 978-3-642-36843-1 e-ISBN 978-3-642-36844-8
DOI 10.1007/978-3-642-36844-8
Springer Heidelberg New York Dordrecht London

Library of Congress Control Number: 2013932548

Printed on acid-free paper

Springer is part of Springer Science+Business Media (www.springer.com)

Preface

The International Workshop on Complex Networks – CompleNet (www.complenet.org) was initially proposed in 2008 with the first workshop taking place in 2009. The initiative was the result of efforts from researchers from the *BioComplex Laboratory in the Department of Computer Sciences at Florida Institute of Technology, USA*, and the from the*Dipartimento di Ingegneria Informatica e delle Telecomunicazioni, Universita' di Catania, Italia*.

CompleNet aims at bringing together researchers and practitioners working on areas related to complex networks. In the past two decades we have been witnessing an exponential increase on the number of publications in this field. From biological systems to computer science, from economic to social systems, complex networks are becoming pervasive in many fields of science. It is this interdisciplinary nature of complex networks that CompleNet aims at addressing. CompleNet 2013 was the fourth event in the series and was hosted by the *Institut fuer Informatik Freie Universitaet Berlin*, Germany from March 13-15, 2013.

This book includes the some of the peer-reviewed works presented at CompleNet 2013. We received 92 submissions from 31 countries. Each submission was reviewed by at least 3 members of the Program Committee. Acceptance was judged based on the relevance to the symposium themes, clarity of presentation, originality and accuracy of results and proposed solutions. After the review process, 10 papers and 8 short papers were selected to be included in this book.

The 18 contributions in this book address many topics related to complex networks and have been organized in seven major groups: (1) Social Networks and Social Media, (2) Opinion and Innovation Diffusion on Networks, (3) Synchronization, (4) Networks & History, (5) Transportation Networks and Mobility, and (6) Theory, Modeling and Metrics of Complex Networks.

We would like to thank the Committee members for the refereeing process work. We deeply appreciate the efforts of our Keynote Speakers: Alex Arenas (Universidad Rovira I Virgili), Martin Warnke (Leuphana Universität Lüneburg) and Thilo Gross (University Of Bristol); their presentations are among the reasons CompleNet 2013 was such a success. We are grateful to our Invited Speakers who enriched CompleNet 2013 with their presentations and insights in the field of Complex

Networks: José F. F. Mendes (University of Aveiro), Ernesto Estrada (University of Strathclyde), Luis R. Izquierdo (University of Burgos), Jesús Gómez-Gardeñes (University of Zaragoza), Maxi San Miguel (Universitat Illes Balears), Albert Diaz-Guilera (Universitat de Barcelona).

Special thanks also go to the Local Organizer and Poster Chair, Markus Luczak-Rösch (Freie Universität Berlin), and the Steering Committee, Ronaldo Menezes (Florida Institute of Technology), Luciano Costa, (University of São Paulo), Giuseppe Mangioni (University of Catania) for their valuable help in organizing CompleNet 2013.

Berlin, Germany
March 2013

Julia Poncela-Casasnovas, Northwestern University
Gourab Ghoshal, Harvard University
Robert Tolksdorf, Freie Universität Berlin

Contents

Network Economics and Transportation

Community Structure and Graph Partitioning

Robustness, Percolation on Networks

Detecting Social Capitalists on Twitter Using Similarity Measures

Nicolas Dugué and Anthony Perez

Abstract. Social networks such as Twitter or Facebook are part of the phenomenon called *Big Data*, a term used to describe very large and complex data sets. To represent these networks, the connections between users can be easily represented using (directed) graphs. In this paper, we are mainly focused on two different aspects of social network analysis. First, our goal is to find an efficient and high-level way to store and process a social network graph, using reasonable computing resources (processor and memory). We believe that this is an important research interest, since it provides a more democratic method to deal with large graphs. Next, we turn our attention to the study of *social capitalists*, a specific kind of users on Twitter. Roughly speaking, such users try to gain visibility by following other users regardless of their contents. Using two similarity measures called *overlap index* and *ratio*, we show that such users may be detected and classified very efficiently.

1 Introduction

Context. In the last decade, large and complex data have been produced through the study of Internet [2], business intelligence [11] or bioinformatics [8]. This phenomenon, known as *Big Data*, raises a lot of research interests. In particular, being able to store, share and analyse such data efficiently constitutes a major research area [9]. In several cases, these data can be represented using graph theory. This is in particular well-suited to study social networks, where connections between users can be easily represented using graphs. Due to a huge increase in the number of users of these social networks, the graphs obtained are very large. Hence, being able to study their structural properties efficiently by using a high-level method and with low computing resources is of important interest. In this paper, we consider

Nicolas Dugué · Anthony Perez
LIFO - Université d'Orléans
e-mail: {nicolas.dugue,anthony.perez}@univ-orleans.fr

G. Ghoshal et al. (Eds.): *Complex Networks IV*, SCI 476, pp. 1–12.
DOI: 10.1007/978-3-642-36844-8_1

the *relation-graph between Twitter users* computed in 2009 [2] and we focus on the behavior of so-called *social capitalists*.

Social Capitalists. These (real) Twitter users (neither bots nor spammers) share a common goal, that is to acquire a maximum number of *followers* to gain visibility. Indeed, the larger the number of followers of an user is, the greatest its influence in the network may be. Besides this obvious interest, accumulating followers is also useful since it has a direct incidence on ranking tweets on search engines [3]. Such a behavior has first been described by Ghosh et al. [3], in a paper considering farm-linking, and especially **spammers**. The authors observed that users that respond the most to request of spammers are mainly *real users*, and they characterize this behavior by calling them *social capitalists*. These users are not healthy for a social network: indeed, since they follow users regardless of their contents, just hoping to be followed back, they help users such as spammers to gain visibility. This specific attitude provides a lot of visibility to their tweets as well, without any content-related reason. Hence, detecting such users is of important interest since it can lead to a better understanding on the *real* influence of a Twitter user.

Our Contribution. The results we provide on this paper are twofold. We first focus on *efficient* and high-level techniques to store and handle very large graphs (several millions vertices and a few billions arcs) without using a large amount of resources[1]. Our experimentations show that regular databases (SQL or NoSQL) are not well-suited to store such graphs. We thus had to consider *graph-oriented databases* to find Dex, a tool which meets our requirements. We next turn our attention to the detection of social capitalists. In particular, we show that social capitalists can be efficiently detected using *similarity measures* on the relation-graph only. An interesting property of our algorithms is that we *do not need* to study the tweets of the users, we only consider the graph topology. For the sake of our study, we use a publicly available graph collected in 2009 (provided by Cha et al. [2]) that contains about 50 millions *anonymized* Twitter users and their relationships (2 billions arcs). More precisely, in order to validate our measures, we consider a *spammer-graph of Twitter*, containing 40000 spammers detected by Ghosh et al. [3] as well as their neighbors, for a total of 15 millions vertices and 1 billion arcs. We will see in Section 3.3 that working on such a graph is well-suited for our purpose. Finally, to validate our detection algorithms, we compare our results to a list of 100000 potential social capitalists detected on the relation-graph using an ad-hoc approach [3]. This approach requires to get a reliable list of spammers, which is really long and may not be easily automated. Furthermore, it returns a list of potential social capitalists which is not exhaustive - only those who follow spammers are found. However, it is important to observe that our algorithms detect most of the users of this list, that have almost their entire neighborhood contained in the spammer-graph (Section 3).

[1] For our studies, we worked on a single machine using *one* core only: AMD Opteron(tm) Processor 6174 800 Mhz 12 cores, with 64 Go RAM.

2 Spammer-Graph

2.1 Definition

Starting from the *relation-graph of Twitter users* $D = (V,A)$, where V represents the set of (*anonymized*) users and $uv \in A$ if and only if user u *follows* user v, we compute the *spammer-graph* $S = (V',A')$. In this case, V' stands for the set of spammers plus their neighbors, and A' the arcs existing between these vertices. To obtain such a graph, we start from a set of 40000 **spammers** provided by Ghosh et al. [3], and then consider their neighbors. Doing so, we preserve about 15 **million** vertices and a bit more than 1 **billion** arcs. Roughly speaking, this represents half the relation-graph of Twitter users provided by Cha et al. [2]. Given $v \in V'$, we define $N^+(v)$ (resp. $N^-(v)$) as the set of *out*- (resp. *in*-) neighbors of v. Consequently, the *out*- (resp. *in*-) degree of a vertex is given by $|N^+(v)|$ (resp. $|N^-(v)|$). From Twitter viewpoint, the in-degree of an user corresponds to the number of users following it, while its out-degree represents the number of users it follows.

2.2 Storage

The main issue with such large-sized graphs is to store them in order to process them efficiently. In their respective works, Cha et al. [2] and Ghosh et al. [3] do not give any details on how they handle the relation graph of Twitter users. In this paper, our first aim is to find a way to store and to process the relation-graph with high-level techniques and without using too many computing resources. The method we suggest can be reproduced easily on a single machine with one processor, the only required resource being a certain amount of memory (about 24 Go to study efficiently a graph with 15 millions vertices and more than 1 billion arcs for instance). We believe that this is fundamental in social network analysis, since it allows researchers from all fields interested in complex networks -computer scientists, biologists, economists, sociologists- to check the provided results and to get their own easily.

To store the graph and analyse it efficiently, we mainly focused on databases. For the sake of completeness, we explore the possibility to use all kinds of databases, that is SQL, NoSQL and graph-oriented. The results of our experimentations show that most of the databases cannot be used to deal with large graphs, even to process simple measures. For instance, MySQL was competitive to load the relation-graph [2], but processing more complex requests like intersection of neighborhoods (a tool needed for our measures) is very slow: several days to process this measure over all the vertices or to index the database efficiently. With Cassandra [5] (NoSQL), even getting the degrees of the vertices from the database could require several hours. Therefore, we tried several graph-oriented databases such as OrientDB [7] or Neo4J [12]. In both cases, we were not able to load the graph in a reasonable amount of time (more than a week was required). Eventually,

Algorithm 1. Extract of the code which processes the similarity measures.

```
1  // Connexion to the database
2  DexConfig cfg = new DexConfig(); Dex d = new Dex(cfg);
3  Database db = d.open(databaseName, true);
4  Session session = db.newSession();
5  Graph g = session.getGraph();
6  // Getting the integers describing the vertices and arcs types
7  int user = g.findType("user");
8  int edge = g.findType("follows");
9  // Getting an object to iterate the collection of vertices
10 Objects objs = g.select(user);
11 ObjectsIterator it = objs.iterator();
12 long v, in, out;
13 // Iterating all the vertices
14 while it.hasNext() do
15     v = it.next();
16     // Getting the out- and in- degrees of the vertex v
17     out = g.degree(v, edge, EdgesDirection.Outgoing);
18     in = g.degree(v, edge, EdgesDirection.Ingoing);
19     // Getting the out- and in- neighbors of the vertex v
20     Objects outVertex = g.neighbors(v, edge, EdgesDirection.Outgoing);
21     Objects inVertex = g.neighbors(v, edge, EdgesDirection.Ingoing);
22     // Processing the similarity measures
23 // Closing the iterators and the database objects is mandatory
24 it.close(); objs.close(); session.close(); db.close(); d.close();
```

Dex [6] appeared as a viable solution for several reasons: high-performance and graph-oriented, a high-level API (see Algorithm 1) and well-documented. On simple demand, we obtained a temporary licence providing us full use of the database. However, if we were able to store the relation-graph in a reasonable amount of time (several hours), our algorithms were not able to run due to some built-in malfunctions. These technical problems -that our study has enlightened- are still under correction. Nevertheless, this issue did not appear when considering the *spammer-graph*, which is smaller than the relation-graph but well-suited for our purpose. On such a graph, both the loading process and our algorithms take only a couple of hours.

Remark. We would like to mention that other ways to store and process the graph would be to use an in-memory adjacency list or a distributed method. Although these low levels techniques should be more efficient, using them require to write tailor-made programs or to have several servers at hand and to set up a complex architecture. Therefore, these methods are not well suited for our purpose, which is to allow all the researchers interested in complex networks to deal with large graphs without high resources nor knowledges.

3 Social Capitalists

3.1 Definition

Similarly to Internet behaviors, where websites administrators perform *links exchange* in order to increase their visibility, some social network users seek to obtain a maximum number of virtual relationships. Twitter is particularly well-suited to observe and study these kinds of behavior. Indeed, microblogging networks are focused on sharing information, not on friendship links. Therefore, it is possible to become visible on this kind of networks by accumulating followers and thus, to spread information efficiently. To achieve this goal, such users (called *social capitalists* in [3]) exploit two relatively straightforward techniques, based on the reciprocation of the *follow* link:

- **FMIFY** (Follow Me and I Follow You): the user ensures his followers that he will follow them back;
- **IFYFM** (I Follow You, Follow Me): at the contrary, these users follow other users hoping to be followed back.

Social capitalists have first been lightened by Ghosh et al. [3] during a study focused on *spammers*. Spammers are engaged in a so-called *farm linking* principle, where the aim is to be followed by a maximum number of users to spread spam links. In [3], the authors observed that users that respond the most to requests of spammers are in fact real users (*i.e.* not spammers nor false accounts). If the fact that such users are not false accounts can be surprising at first sight, it can actually be easily explained: as mentioned previously, social capitalists use the **FMIFY/IFYFM** principles, and follow back users that follow them regardless of the content of their tweets.

3.2 Similarity Measures

In order to detect social capitalists, we make use of two similarity measures, namely *overlap index* (introduced in [10]) and *ratio*. As we shall see, the first one will enable us to detect potential social capitalists, while the latter one will allow us to determine if a given user uses the **FMIFY** or **IFYFM** principle.

Definition 1. Given two sets A and B, the *overlap index* of A and B (which is between 0 and 1) is given by:

$$O(A,B) = \frac{|A \cap B|}{\min\{|A|,|B|\}}$$

For every vertex v, we first apply the overlap index on the sets $A = N^+(v)$ and $B = N^-(v)$. We claim that this allows us to detect users that are likely to be social capitalists. Indeed, due to the applied principles, the relation between their in- and out-neighbors must be strong, and thus most of the users they follow should follow them back. In particular, this means that their overlap index must be close to 1. We next use Definition 2 to classify more precisely such users.

Definition 2 (Ratio). Let $D = (V,A)$ be any directed graph, and $v \in V$ be any vertex. The *ratio* of v is given by:

$$R(v) = \frac{|N^+(v)|}{|N^-(v)|}$$

Intuitively, social capitalists following the **IFYFM** principle should have a ratio greater than 1 (*i.e.* more followers than followers), while the users following the **FMIFY** should have a ratio less than 1. Recall that, in both cases, the expected ratio should be close to 1. However, as we shall see in Section 3.6, another behavior arises that we call *passive* (see Figure 1). Unlike other social capitalists (called *active*), these users find their in-degree large enough, and thus estimate having enough influence. Once they have reached this point, they stop using the aforementioned principles but still get more and more followers. Consequently, their overlap index should still be close to 1 whereas their ratio should be a lot smaller than 1.

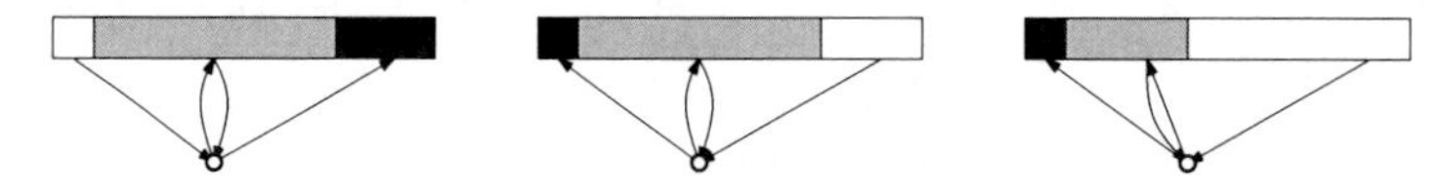

Fig. 1 Different behaviors of social capitalists: on the left, the **FMIFY** principle, with a set of users waiting to be followed back, in the middle, the **IFYFM** principle, the social capitalist being waiting for a follow back, and on the right a **passive FMIFY** user

3.3 Detection in the Spammer-Graph

We claim that detecting social capitalists in the spammer-graph can provide useful and reliable information about social capitalists in the whole relation-graph. To validate this, we use a list of 100000 users considered as potential social capitalists in the relation-graph by Ghosh et al. [3]. These users have been detected through an ad-hoc approach, namely detecting users that respond the most to requests of spammers. Therefore, these 100000 users appear in the spammer-graph. More precisely, as indicated Tables 1 and 2, most of these users contain almost their entire neighborhood in the spammer-graph. To process the degree of every vertex in the relation graph of Twitter, we used the file provided by Cha et al. [2] (where every row corresponds to an arc of the graph) as well as a MySQL database[2].

This means in particular that most users from the list can safely be considered as social capitalists in the spammer-graph. Hence, we will use such a list to measure the efficiency of our detection algorithms. As we shall see in Section 3.6, we detect a large majority of these users. These observations allow us to validate the accuracy of our results, and to carry on the study of social capitalists on the spammer-graph. Furthermore, we will see that most of the users detected by our algorithms also have a large part of their neighborhoods included in the spammer-graph.

[2] Recall that we were not able to efficiently analyse the relation-graph on Dex.

Table 1 Average in- and out-degree of the 100000 potential social capitalists detected by [3] in both the relation and the spammer graphs (column with "-spam"). The column LIMIT show the number of social capitalists (sorted by decreasing in-degree) used.

LIMIT	avg(in)	avg-spam(in)
100000	3313	3059
50000	5786	5316
10000	16993	15100

LIMIT	avg(out)	avg-spam(out)
100000	3455	3222
50000	5633	5292
10000	15397	14333

Table 2 Number of potential social capitalists from the list [3] having % of their in and out-neighbors in the spammer graph

%	in-	out-	in- and out-
90	88407	73431	72368
80	95107	90359	83860
70	96901	96047	89436

3.4 Threshold

As stated in Section 3.2, users with an overlap index close to 1 are likely to be social capitalists. We now define what *close to* 1 means. To that aim, we use the list of 100000 users considered as potential social capitalists in the relation-graph [3]. More precisely, for the sake of our study, we choose to consider the 72368 users that have 90% of their in- and out-neighbors in the spammer graph (see Figure 2). We compute their overlap index in the spammer-graph as well as the distribution of the values of the overlap index among them.

We observe that a few users of this list have an overlap less than 0.6. We may wonder whether these users really are social capitalists. Actually, the only conclusion we can draw is that they do not use the principles described before. Above 0.6, there is a dramatic increase in the number of users for each interval. Notice that 80% of these users have an overlap greater than 0.8. Since we want to preserve users who are the most likely to be social capitalists and to avoid too many false positives, we must use a high threshold. Due to these observations, we choose 0.8.

Because these measures are computed on the spammer-graph where these potential social capitalists have at least 90% of their in- and out-neighborhood, the threshold should be different in the *relation-graph of Twitter users*. Indeed, using Definition 1, we can see that any user with less than 10% of its neighborhood outside the spammer-graph has an overlap index in the relation-graph greater than 0.9 times its overlap index in the spammer-graph. Hence, if we choose 0.8 as a threshold in the spammer-graph, we should use 0.72 in the relation-graph.

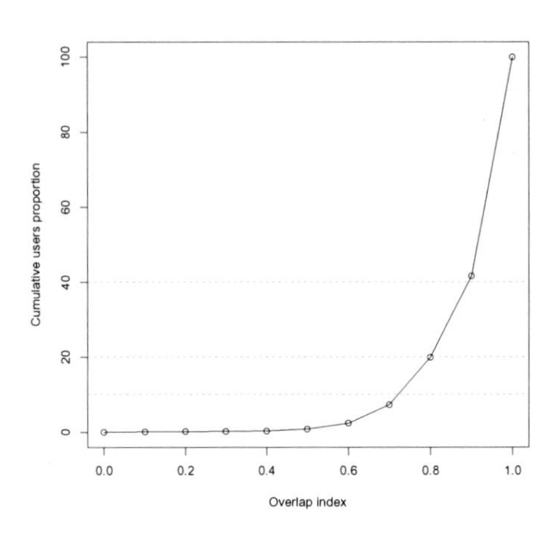

Fig. 2 Overlap index of the in and out neighborhoods of 72368 potential social capitalists from the list [3]

3.5 Examples

To give a better picture on social capitalists, we provide a short list of well-known social capitalists such as Barack Obama, Britney Spears or JetBlue Airways [3].

Table 3 Well-known or obvious social capitalists, their overlap and ratio obtained in the relation-graph by using the Twitter API. All the values are rounded.

screen name	name	followers	followees	overlap	ratio
IFOLLOWBACKJP	TFBJP	$1.2 \cdot 10^5$	$1.1 \cdot 10^5$	0.97	0.92
itsrealchris	iFollowBack	$1.7 \cdot 10^5$	$1.6 \cdot 10^5$	0.81	0.94
AllFollowMax	TFBJP	$4.2 \cdot 10^4$	$4.3 \cdot 10^4$	0.99	1.04
BarackObama	Barack Obama	$2.5 \cdot 10^7$	$6.7 \cdot 10^5$	0.77	0.03
britneyspears	Britney Spears	$2.2 \cdot 10^7$	$4.1 \cdot 10^5$	0.82	0.02
JetBlue	JetBlue Airways	$1.7 \cdot 10^6$	$1.1 \cdot 10^5$	0.74	0.06
Starbucks	Starbucks Coffee	$3.2 \cdot 10^6$	$7.9 \cdot 10^4$	0.77	0.02

We want to mention that TBJP means *Team Follow Back Japan*. Using our similarity measures, we can detect these users because their overlap index is greater than 0.72, the threshold in the relation-graph. The first two users are considered as **FMIFY** users because their ratio is less than 1. The third one is classified as a **IFYFM** with a ratio greater than 1. We observe that the ratio of the four last ones is really close to 0, implying that they are passive social capitalists, which makes sense since they correspond to famous Twitter accounts.

3.6 Experimental Results

We now present the experimental results we obtained on the spammer-graph. We considered users having an overlap index greater than the threshold 0.8 chosen before. The column *vertices* presents vertices observed w.r.t. to the in-degree (*in*) constraint, while the column % corresponds to the percentage of these vertices for a given constraint *ratio*. We mainly consider users with high in-degree since they correspond to users that used the **FMIFY/IFYFM** principles successfully.

Table 4 Detecting social capitalists on the spammer-graph

in	vertices	ratio	%
> 500	137267		
		> 1	63
		$[0.7;1]$	24
> 2000	40725		
		> 1	61
		$[0.7;1]$	28
> 10000	5344		
		> 1	63
		$[0.7;1]$	23
		< 0.7	14

Interpreting the Results. When the in-degree is greater than 500, we observe the two behaviors **IFYFM** (63% of the users have a ratio greater than 1) and **FMIFY** (24% of the users have a ratio between 0.7 and 1 and a large overlap index). Similarly, these two groups are enhanced when we consider users with in-degree greater than 2000 and 10000. On the last row of Table 4, we also notice that *passive* social capitalists arise: 14% of the users have a ratio less than 0.7. In average, these users have an in-degree around 20 times greater than their out-degree.

Validating the Results. To confirm our observations, we make use of the list of 100000 potential social capitalists detected by Ghosh et al. [3]. Since the neighborhood of these users is almost entirely contained in the spammer-graph (Section 3.3), they should be detected by our algorithms. We would like to mention that about 12000 vertices of this list (12%) have an in- or out-degree less than 500, explaining why they are not detected in the measures we present.

Observe that we detect many social capitalists which are not in the list. This comes from the fact that it is made of users which follow spammers, and is therefore not exhaustive. This is especially true considering the last row of Table 5. It corresponds to the detection of passive social capitalists, which are not applying the **FYIFM/IFYFM** principles anymore and so, stopped responding to spammers.

Table 5 The percentage of social capitalists from the list provided in [3] that are detected by our algorithms. We detect 67% of this list, but more than 12% of the vertices of the list have a degree less than 500, so, precluding these users, this represents a recall of 76%.

in	ratio	detected	list [3]
> 500	> 1	86594	54870
> 500	$[0.7; 1]$	33291	12347
> 2000	> 1	24848	23060
> 2000	$[0.7; 1]$	11649	8093
> 10000	> 1	3378	3260
> 10000	$[0.7; 1]$	1232	1188
> 10000	< 0.7	734	79

Table 6 The percentage of detected social capitalists having at least 90% of their in- and out-neighbors in the spammer-graph

in	ratio	detected	90%-in	90%-out	90%-in and out
> 500	> 1	86594	71984	60690	59585
> 500	$[0.7; 1]$	33291	22331	23288	21691
> 2000	> 1	24848	23374	22632	22583
> 2000	$[0.7; 1]$	11649	9637	9590	9412
> 10000	> 1	3378	3268	3219	3217
> 10000	$[0.7; 1]$	1232	1058	1044	1037
> 10000	< 0.7	734	45	200	37

To investigate further the coherence of our results, we show that the social capitalists detected by our algorithms also have their neighborhood almost entirely contained in the spammer-graph. More precisely, we count the number of detected social capitalists that contain at least 90% of their neighborhood in the spammer-graph (Table 6). Recall that *any* user detected by our algorithms containing at least 90% of both its in- and out-neighborhood in the spammer-graph will have an overlap of *at least* 0.72 in the relation graph of Twitter.

Once again, we observe a difference if we consider users with large overlap, large in-degree but small ratio. This follows from the fact that such users are most likely to be *passive* social capitalists, and hence have a lot of followers outside the spammer-graph.

4 Conclusion and Perspectives

In this paper, we focused on efficiently storing and analysing the relation-graph between Twitter users. In particular, we focused on handling really large graphs (several millions vertices and a few billions arcs) using reasonable computing resources. We tried several graph-oriented databases, and noticed that most of them do not seem well-suited for storing such large graphs. However, using Dex [6], we were able to store a graph containing about 15 million vertices and 1 billion arcs. This so-called spammer-graph allowed us to develop a study on *social capitalists*, a particular kind of users in Twitter that tend to follow everybody that follows them regardless of their contents. We focused on the detection of such users, using similarity measures called *overlap index* and *ratio*. In particular, by comparing our results with a list of social capitalists provided by Ghosh et al. [3], we observed that our detection algorithm provides an efficient way to detect social capitalists.

We believe that the study of social capitalists as we initiated it could lead to a significant amount of new research. In particular, we would like to study more precisely the behavior of social capitalists to see if they make use of more elaborate strategies to gain visibility. Another perspective that arises from our work is to create several false Twitter accounts applying the aforementioned principles used by social capitalists. More precisely, one interesting study would be to follow users applying the **FMIFY** principle, and analyse the impact on the influence of the account. This would allow us to validate the efficiency of such strategies, depending on the credibility of the created accounts. We have actually started such a study, using non-anonymized data provided by Kwak et al. [4]. We also plan to check if social capitalists are present on other social networks such as LiveJournal or Google+. Finally, we would like to study the impact of social capitalists on the community detection on the Twitter graph. Since social capitalists have strong connections with their neighborhoods, and have high degree in average, they may have specific roles in communities. To do so, we need to use community detection algorithms that can handle a graph as large as the Twitter graph, like the one described in [1].

Acknowledgement. Nicolas Dugué benefits from a PhD grant of the *Conseil Général du Loiret*. This work is partly supported by the INEX project (`http://traclifo.univ-orleans.fr/INEX`) funded by the *Conseil Général du Loiret*. The authors thank Sylvain Jubertie and the anonymous referees for their helpful comments that improved the presentation of this paper.

References

1. Blondel, V.D., Guillaume, J.L., Lambiotte, R., Lefebvre, E.: Fast unfolding of communities in large networks. J. of Stat. Mech.: Theory and Experiment 2008(10), 10,008 (2008)
2. Cha, M., Haddadi, H., Benevenuto, F., Gummadi, K.P.: Measuring User Influence in Twitter: The Million Follower Fallacy. In: ICWSM 2010: Proc. of int. AAAI Conference on Weblogs and Social (2010)
3. Ghosh, S., Viswanath, B., Kooti, F., Sharma, N.K., Korlam, G., Benevenuto, F., Ganguly, N., Gummadi, K.P.: Understanding and Combating Link Farming in the Twitter Social Network. In: Proc. of the 21st Int. Conference on World Wide Web, WWW 2012, pp. 61–70 (2012)

4. Kwak, H., Lee, C., Park, H., Moon, S.: What is Twitter, a social network or a news media? In: Proc. of the 19th Int. Conference on World Wide Web, WWW 2010, pp. 591–600 (2010)
5. Lakshman, A., Malik, P.: Cassandra: a structured storage system on a p2p network. In: Proc. of the 28th ACM Symp. on Princ. of Distributed Comput., PODC 2009, p. 5 (2009)
6. Martínez-Bazan, N., Águila Lorente, M.A., Muntés-Mulero, V., Dominguez-Sal, D., Gómez-Villamor, S., Larriba-Pey, J.L.: Efficient Graph Management Based On Bitmap Indices. In: Proc. of the 16th Int. Database Eng. & Appl. Symp., IDEAS 2012, pp. 110–119 (2012)
7. OrientDB (1999), http://www.orientdb.org/
8. Schatz, M.C., Langmead, B., Salzberg, S.L.: Cloud computing and the DNA data race. Nat. Biotech. 28(7), 691–693 (2010)
9. Schuett, T., Pierre, G.: ConpaaS, an integrated cloud environment for big data. ERCIM News 2012(89) (2012)
10. Simpson, G.G.: Mammals and the nature of continents. Am. J. of Science (241), 1–41 (1943)
11. Thusoo, A., Sarma, J.S., Jain, N., Shao, Z., Chakka, P., Zhang, N., Anthony, S., Liu, H., Murthy, R.: Hive - a petabyte scale data warehouse using hadoop. In: IEEE 26th Int. Conference on Data Eng., pp. 996–1005 (2010)
12. Vicknair, C., Macias, M., Zhao, Z., Nan, X., Chen, Y., Wilkins, D.: A comparison of a graph database and a relational database: a data provenance perspective. In: Proc. of the 48th Annu. Southeast Reg. Conference, ACM SE, pp. 42:1–42:6 (2010)

Semantics of User Interaction in Social Media

Folke Mitzlaff, Martin Atzmueller, Gerd Stumme, and Andreas Hotho

Abstract. In ubiquitous and social web applications, there are different user traces, for example, produced explicitly by "tweeting" via twitter or implicitly, when the corresponding activities are logged within the application's internal databases and log files.

For each of these systems, the sets of user interactions can be mapped to a network, with links between users according to their observed interactions. This gives rise to a number of questions: Are these networks independent, do they give rise to a notion of user relatedness, is there an intuitively defined relation among users?

In this paper, we analyze correlations among different interaction networks among users within different systems. To address the questions of interrelationship between different networks, we collect for every user certain external properties which are independent of the given network structure. Based on these properties, we then calculate semantically grounded reference relations among users and present a framework for capturing semantics of user relations. The experiments are performed using different interaction networks from the twitter, flickr and BibSonomy systems.

1 Introduction

With the increasing availability of mobile internet connections, social applications are ubiquitously integrated in the user's daily life. By interacting with such systems, the user is leaving traces within the different databases and log files, e. g., by updating the current status via twitter or chatting with social acquaintances via facebook.

Folke Mitzlaff · Martin Atzmueller · Gerd Stumme
Knowledge and Data Engineering Group, University of Kassel,
Wilhelmshöher Allee 73, D-34121 Kassel, Germany
e-mail: {mitzlaff,atzmueller,stumme}@cs.uni-kassel.de

Andreas Hotho
Data Mining and Information Retrieval Group, University of Würzburg,
Am Hubland, D-97074 Würzburg, Germany
e-mail: hotho@informatik.uni-wuerzburg.de

G. Ghoshal et al. (Eds.): *Complex Networks IV*, SCI 476, pp. 13–25.
DOI: 10.1007/978-3-642-36844-8_2

Ultimately, each type of such traces gives rise to a corresponding network of user relatedness, where users are connected if they interacted either explicitly (e. g., by establishing a "friendship" link within in an online social network) or implicitly (e. g., by visiting a user's profile page). We consider a link within such a network as evidence for user relatedness and call it accordingly *evidence network* or *interaction network*.

These interaction networks are of large interest for many applications, such as recommending contacts in online social networks or for identifying groups of related users [19]. Nevertheless, it is not clear, whether every such interaction network captures meaningful notions of relatedness and what the semantics of different aggregation levels really are.

As multifaceted humans are, as many reasons for individuals being related exists. Ultimately, it is therefore not possible to judge whether an interaction network is "meaningful" or not. Nevertheless, certain networks are more probable than others and give rise to more traceable notions of relatedness. In this paper, we compare different networks based on external notions of relationship between users, especially geographical proximity. We argue that the geographical proximity of two users is a prior for other important notions of relationship, such as common language and cultural background. We therefore use such an external measure of relatedness as a proxy for semantically grounded relationship among users.

This paper proposes an experimental methodology for assessing the semantics of evidence networks and similarity metrics therein. The contributions of the paper can be summarized as follows: The presented methodology is applied to a broad range of evidence networks obtained from twitter, flickr and BibSonomy as well as different similarity metrics for calculating similarity of nodes within a network. The obtained results thus yield a *semantic grounding* of evidence networks and similarity metrics, which are merely based on structural properties of the networks. Furthermore, we consider both established reference sources such as tagging data, as well as geographical locational data as a proxy for semantic relatednesss. Finally, the collected analysis results, and especially the proposed methodology can serve as a foundation for further applications, such as *community mining* and *link prediction* tasks.

The remainder of this paper is structured as follows: First, we discuss related work in Section 2. In Section 3 we present all considered network data. Next, Section 4 introduces the proposed methodology for assessing semantics of user relatedness in evidence networks and presents according experimental results obtained from the considered networks. In Section 5 we summarize the results and point at future directions of research and applications for the results presented in this paper.

2 Related Work

The present paper tackles the problem of grounding the semantics of user relatedness, induced by online social networks: It combines approaches from the field of *social network analysis* and methods for measuring *distributional similarity*. Furthermore, it also considers *geospatial analysis* in online social media. Notably, the present work is based on [21], where semantics of Wikipedia based co-occurrence networks of named entities are analyzed with respect to category assignments to corresponding articles in Wiktionary as well as corresponding geo-location.

The field of distributional similarity and semantic relatedness has attracted a lot of attention in literature during the past decades (see [3] for a review). Several statistical measures for assessing the similarity of words are proposed, e. g., in [13,4,6,11,25], especially in the context of social bookmarking systems [2].

The task of calculating similarity between individuals within a social network is closely related to the *link prediction* and *user recommendation* tasks, where "missing" links are predicted based on the network structure. In the context of social networks, the task of predicting (future) links is especially relevant for online social networks, where social interaction is significantly stimulated by suggesting people as contacts which the user might know. From a methodological point of view, most approaches build on different similarity metrics on pairs of nodes within weighted or unweighted graphs [7, 12, 16, 17]. A good comparative evaluation of different similarity metrics is presented in [15]. In [26] a topic sensitive user ranking in the context of online social media is proposed.

The analysis of online social media, the interrelations of the involved actors, and the involved geospatial extents have attracted a lot of attention during the last decades, especially for the microblogging system twitter. A thorough analysis of fundamental network properties and interaction patterns in twitter can be found in [10].

An analysis of a location-based social network with respect to the user attributes is investigated in [14]. Interdependencies of social links and geospatial proximity are investigated in [24, 18, 9, 23, 8], especially concerning the correlation of the probability of friendship links and the geographic distance of the corresponding users. The impact of subgroups of users and communities is analyzed in [1]. In [20], interaction networks which accrue as aggregations of log files within the social tagging system BibSonomy are introduced and analyzed.

In contrast to previous work, the present work focuses on the question, whether a given social network gives rise to a notion of relatedness among its nodes and how different network variants, such as directedness and edge weights have an impact on the resulting network semantics. The proposed methodology is applied to different networks and structural similarity metrics, giving new insights into the semantics of those networks and their variants as well as the considered similarity metrics.

3 Network Data

Evidence Networks in BibSonomy. BibSonomy is a social bookmarking system where users manage their bookmarks and publication references via *tag* annotations (i. e., freely chosen keywords). Most bookmarking systems incorporate additional relations on users such as "*my network*" in del.icio.us and "*friends*" in BibSonomy.

But beside those explicit relations among users, different relations are established implicitly by user interactions within the systems, e. g., when user u looks at user v's resources. As all of BibSonomy's log files were accessible, a broad range of interaction networks was available. In particular, we considered the directed *Friend-Graph*, containing an edge (u, v) iff user u has added user v as a friend, the directed *Copy-Graph* which contains an edge (u, v) with weight $c \in \mathbb{N}$, iff user u has copied c resources, i. e., a publication reference from user v and the directed *Visit-Graph*, containing an edge (u, v) with label $c \in \mathbb{N}$ iff user u has navigated c times to the user page of user v.

Evidence Networks in twitter. We also considered the microblogging service twitter. Using twitter, each user publishes short text messages (called "*tweets*") which may contain freely chosen *hashtags*, i. e., distinguished words which are used for marking keywords or topics. Furthermore, users may "cite" each other by "retweeting": A user u retweets user v's content, if u publishes a text message containing "RT @v:" followed by (an excerpt of) v's corresponding tweet. Users may also explicitly follow other user's tweets by establishing a corresponding friendship-like link. For our analysis, we considered the directed *Follower-Graph*, containing an edge (u, v) iff user u follows the tweets of user v and the *ReTweet-Graph*, containing an edge (u, v) with label $c \in \mathbb{N}$ iff user u cited (or "retweeted") exactly c of user v's tweets.

Evidence Networks in flickr. The flickr system focuses on organizing and sharing photographs collaboratively. Users mainly upload images and assign arbitrary tags but also interact, e. g., by establishing contacts or commenting on other users images. For our analysis we extracted the directed *Contact-Graph*, containing an edge (u, v) iff user u added user v to its personal contact list, the directed *Favorite-Graph*, containing an edge (u, v) with label $c \in \mathbb{N}$ iff user u added exactly c of v's images to its personal list of favorite images as well as the directed *Comment-Graph*, containing edge (u, v) with label $c \in \mathbb{N}$ iff user u posted exactly c comments on v's images.

General Structural Properties. Table 1 summarizes major graph level statistics for the considered networks which range in size from thousands of edges (e. g., the Friend-Graph) to more than one hundred million edges (flickr's Contact-Graph). All networks obtained from BibSonomy are complete and therefore not biased by a previous crawling process. In return, effects induced by limited network sizes have to be considered.

4 Analysis of Network Semantics

In Section 3, we introduced various explicit and implicit interaction networks from different applications. In this section, we tackle the problem of assessing the "meaning" of relations among pairs of vertices within such a network. This analysis then

Table 1 High level statistics for all networks with density d, the number of strongly connected components #scc and the size of the largest strongly connected component SCC

	$\lvert V_i \rvert$	$\lvert E_i \rvert$	d	#scc	SCC
Copy	$1,427$	$4,144$	$2 \cdot 10^{-3}$	$1,108$	309
Visit	$3,381$	$8,214$	10^{-3}	$2,599$	717
Friend	700	$1,012$	$2 \cdot 10^{-3}$	515	17
ReTweet	$826,104$	$2,286,416$	$3,4 \cdot 10^{-6}$	$699,067$	$123,055$
Follower	$1,486,403$	$72,590,619$	$3,3 \cdot 10^{-5}$	$198,883$	$1,284,201$
Comment	$525,902$	$3,817,626$	$1,4 \cdot 10^{-5}$	$472,232$	$53,359$
Favorite	$1,381,812$	$20,206,779$	$1,1 \cdot 10^{-5}$	$1,305,350$	$76,423$
Contact	$5,542,705$	$119,061,843$	$3,9 \cdot 10^{-6}$	$4,820,219$	$722,327$

gives insights into the question, whether and to which extent the networks give rise to a common notion of *semantic relatedness* among the contained vertices. For this, we apply an experimental methodology, which was previously used for assessing semantical relationships within co-occurrence networks [21]. The basic idea is simple: We consider well founded notions of relatedness, which are naturally induced by external properties of the corresponding vertex sets, as, e. g., similarity of the applied tag assignments in BibSonomy or geographical distance between users in twitter. We than compute for each pair of vertices within a network these "semantic" similarity metrics and correlate them with different measures of structural similarity in the considered network.

4.1 Vertex Similarities

Below, we apply two well-established similarity functions in corresponding unweighted variants, namely the cosine similarity COS and the Jaccard Index JC as well as the corresponding weighted variants $\widetilde{COS}$ and $\widetilde{JC}$, following the presentation in [22].

Additionally we apply a modification of the *preferential PageRank* which we adopted from our previous work on folksonomies [5]: For a column stochastic adjacency matrix A and damping factor α, the *global* PageRank vector $\mathbf{w}$ with uniform *preference vector* $\mathbf{p}$ is given as the fixpoint of $\mathbf{w} = \alpha A\mathbf{w} + (1-\alpha)\mathbf{p}$. In case of the *preferential PageRank* for a given node i, only the corresponding component of the preference vector is set. For vertices x, y we set accordingly $PPR(x,y) := \mathbf{w}_{(x)}[y]$, that is, we compute the preferential PageRank vector $\mathbf{w}_{(x)}$ for node x and take its y'th component. We calculate the adopted preferential PageRank score by subtracting the global PageRank score PR from the preferential PageRank score in order to reduce frequency effects and set

$$PPR{+}(x,y) := PPR(x,y) - PR(x,y).$$

4.2 Semantic Reference Relations

For assessing the semantic similarity of two nodes within a network, we look for external properties which give rise to a well founded notion of relatedness. In the following, we consider the similarity of users based on the applied tags in BibSonomy and flickr, as well as the applied hashtags in twitter. We also consider geographical distance of users in twitter and flickr.

Tag Similarity. In the context of social tagging systems like BibSonomy, the cosine similarity is often used for measuring semantic relatedness (see, e. g., [2]).

We compute the cosine similarity in the vector space $\mathbb{R}^T$, where, for user u, the entries of the vector $(u_1, \ldots, u_T) \in \mathbb{R}^T$ are defined by $u_t := w(u,t)$ for tags t where $w(u,t)$ is the number of times user u has used tag t to tag one of her resources (in case of BibSonomy and flickr) or the number of times user u has used hash tag t in one of her tweets (in case of twitter).

Geographical Distance. In twitter and flickr, users may provide an arbitrary text for describing his or her location. Accordingly, these location strings may either denote a place by its geographic coordinates, a semi structured place name (e. g., "San Francisco, US"), a colloquial place name (e. g., "Motor City" for Detroit) or just a fantasy name. Also the inherent ambiguity of place names (consider, e. g., "Springfield, US") renders the task of *exactly* determining the place of a user impossible. Nevertheless, by applying best matching approaches, we assume that geographic locations can be determined up to a given uncertainty and that significant tendencies can be observed by averaging over many observations.

We used Yahoo!'s Placemaker™ API[1] for matching user provided location strings to geographic locations with automatic place disambiguation. In case of flickr, we obtained geographic locations for $320,849$ users and in case of twitter for $294,668$ users. Geographical distance of users is then simply given by the distance of the centroids for the correspondingly matched places.

Please note that geographic distance correlates with many secondary notions of relatedness between users, such as, e. g., language, cultural background and habits.

4.3 Grounding of Shortest Path Distance

For analyzing the interdependence of *semantic* and *structural* similarity between users, we firstly consider a very basic measure of structural relatedness between two nodes in a network, namely their respective shortest path distance. We ask, whether users which are direct neighbors in an evidence network tend to be more similar than distant users. That is, for every shortest path distance d and every pair of nodes u, v with a shortest path distance d, we calculated the average corresponding similarity scores $COS(u,v)$, $JC(u,v)$, $PPR(u,v)$ with variants and geographic distance. To

[1] http://developer.yahoo.com/geo/placemaker/

rule out statistical effects, we repeated for each network G the same calculations on shuffled null model graphs.

Semantic Similarity. Figure 1 shows the resulting plots for each considered network separately. Though the obtained average similarity scores vary greatly in magnitude for different networks (e. g., a maximum of 0.22 for the Friend-Graph in BibSonomy compared to a maximum of 0.1 for the Visit-Graph), they also share a common pattern: Direct neighbors are in average significantly more similar than distant pairs of users. And with a distance of two to three, users tend to be less similar than in average (in case of the ReTweet graph, users are more similar than in average up to a distance of eight). For the Visit-Graph, the Comment-Graph, the Follower-Graph and the ReTweet graph, the average similarity scores approach the global average similarity again. For distances around a network's diameter, the number of observations is too small, resulting in less pronounced tendencies for very distant nodes.

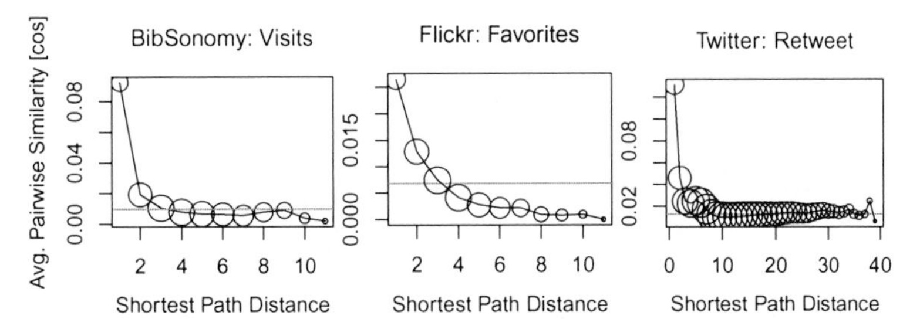

Fig. 1 Average pairwise cosine similarity based on the users' tag assignments relative to the shortest path distance in the respective networks where the global average is depicted in gray and the point size scales logarithmically with the number of pairs.

Geographic Distance. For average geographic distances of users in flickr and twitter, we repeated the same calculations, as depicted in Figure 2. Firstly, we note the overall tendency, that direct neighbors tend to be located more closely than distant pairs of users within a network. Additionally, the average geographic distance of users then approaches the global average, and increases again after a certain plateau. As for the ReTweet-Graph, the average geographic distance remains at the global average level, once reached at a shortest path distance of ten.

*Discussion.*It is worth emphasizing, that in all considered evidence networks, the relative position of users already gives rise to a semantically grounded notion of relatedness, even in case of implicit networks, which are merely aggregated from usage logs as, e. g., the Visit-Graph. But one has to keep in mind that all observed tendencies are the result of averaging over a very large number of observations (e. g., $34,282,803,978$ pairs of nodes at distance four in the Follower-Graph). Therefore, we cannot deduce geographic proximity from topological proximity for a given pair of users, as even direct neighbors in the Follower-Graph are in average located

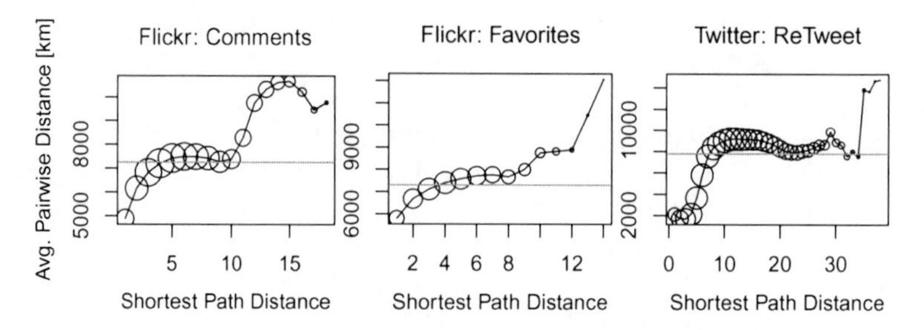

Fig. 2 Shortest path distance vs. average pairwise geographic distance in flickr. The global average is depicted in gray and the point size scales logarithmically with the number of pairs.

$4,000$ kilometers apart from each other. But the proposed analysis aims at revealing semantic tendencies within a network and for comparing different networks (e. g., the Retweet-Graph better captures geographic proximity of direct neighbors in the graph). The experimental setup also allows to assess the impact of certain network variations, such as weighted and unweighted or directed and undirected networks, as exemplified in Section 4.5.

4.4 Grounding of Structural Similarity

We now turn our focus towards different measures of structural similarity for nodes within a given network. There is a broad literature on such similarity metrics for various applications, such as link prediction [15] and distributional semantics [6,21]. We thus extend the question under consideration in Section 4.3, and ask, which measure of structural similarity best captures a given semantically grounded notion of relatedness among users. In the scope of the present work, we consider the cosine similarity and Jaccard index, which are based only on the direct neighborhood of a node as well as the (adjusted) preferential PageRank similarity which is based on the whole graph structure (refer to Section 4.1 for details).

Ultimately, we want to visualize correlations among structural similarity in a network and semantic similarity, based on external properties of nodes within it. We consider, again, semantical similarity based on users' tag assignments in BibSonomy, flickr and hash tag usage in twitter as well as geographic distance of users in flickr and twitter. In detail: For a given network $G = (V, E)$ and structural similarity metric S, we calculate for every pair of vertices $u, v \in V$ their structural similarity $S(u, v)$ in G as well as their semantic similarity and geographic distance. For visualizing correlations, we create plots with structural similarity at the x-axis and semantic similarity at the y-axis. As plotting the raw data points is computationally infeasible (in case of the Contact-Graph $30,721,580,000,000$ data points), we binned the x-axis and calculated average semantical similarity scores per bin. As the distribution of structural similarity scores is highly skewed towards lower similarity scores (most pairs of nodes have very low similarity scores), we applied logarithmic binning, that is, for a structural similarity score $x \in [0, 1]$ we determined the corresponding bin via $\lfloor \log(x \cdot b^N) \rfloor$ for given number of bins N and suitable base b. Pragmatically, we determined the base relative to the machine's floating point precision ϵ resulting in $b := \epsilon^{\frac{-1}{N}}$.

Semantic Similarity. Figure 3 shows the obtained results for each considered network separately. We firstly note, that the cosine similarity metric and the Jaccard index are highly correlated. Secondly, the adjusted preferential PageRank similarity consistently outperforms the other similarity metrics with respect to magnitude and monotonicity (except for BibSonomy's Friend-Graph and flickr's Contact-Graph).

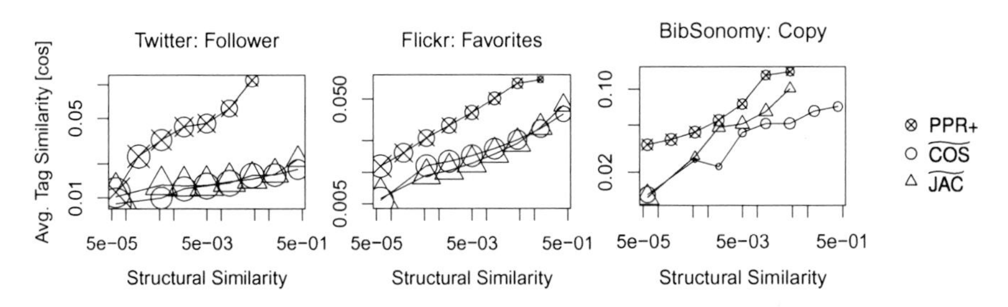

Fig. 3 Average pairwise semantic similarity based on tags users assigned to resources in BibSonomy and flickr or hash tag usage in twitter, relative to different structural similarity scores in the corresponding networks. The point size scales logarithmically with the number of pairs.

Geographic Distance. As for geographic distances, Figure 4 shows the observed correlations for structural similarity in the different evidence networks and the corresponding average pairwise distance. In all but flickr's Favorite-Graph, for both local neighborhood based similarity metrics COS and JC, the average distance first decreases, but then increases again. This behavior is most pronounced in twitter's ReTweet-Graph. In the Favorite-Graph, both COS and JC monotonically decrease with increasing similarity score. On the other hand, the average distance decreases monotonically with increasing preferential PageRank score PPR consistently in all considered networks, except the ReTweet-Graph, where the average distance stays at a level of around 2.000 kilometers for similarity scores > 0. Generally (except for the ReTweet-Graph), it yields average distance values which are magnitudes below those obtained via the local similarity metrics.

Discussion. Again, the obtained results only point at tendencies of the considered similarity metrics in capturing geographic proximity by means of structural similarity. Nevertheless, the adjusted preferential PageRank similarity consistently outperforms the other considered metrics. We therefore conclude that from all considered similarity metrics, the adjusted preferential PageRank similarity best captures the notion of geographic proximity. This is especially of interest, as the geographic proximity is a prior for many properties users may have in common, such as, e. g., language, cultural background or habits. twitter's ReTweet-Graph seems to

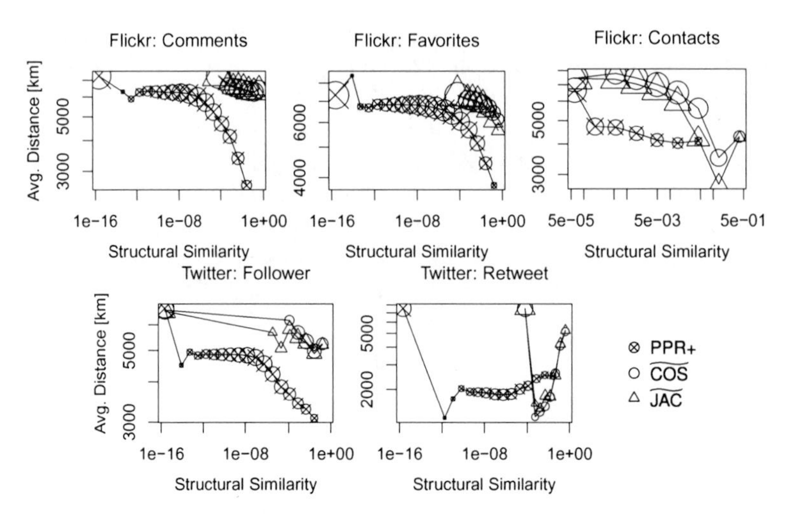

Fig. 4 Average pairwise distance relative to different structural similarity scores in the corresponding networks. The point size scales logarithmically with the number of pairs.

encompass the strongest geographic binding, as indicated in the relative low average distance for direct neighbors (cf. Figure 2 and the overall low average distance for higher preferential PageRank similarity scores (cf. Figure 4). Of course, other established similarity metrics (e. g., [12, 7, 6]) can be applied as well and are the subject of future considerations.

4.5 Case Study: BibSonomy

Most social networks are very sparse and in case of directed networks, dropping the direction of edges is a way of increasing a network's density. This might be of interest, e. g., for calculating a similarity function like $\widetilde{COS}$, which is based on a node's neighborhood. The rational would be, that with a more dense adjacency matrix, more non-zero similarity scores are obtained.

The Impact of Directions. We apply the semantic correlation analysis from Section 4.4 to assess the impact of dropping edge directions in the considered evidence networks in BibSonomy.

Figure 5 shows the corresponding plots, where the average similarity scores for the corresponding undirected networks are depicted in gray. The impact of dropping the directedness varies greatly among the different networks and similarity metrics. Firstly, the semantics of the cosine similarity in the Visit-Graph changes dramatically, by showing negative correlations with the (average) semantic similarity score in case of the undirected network. In the other networks, the cosine similarity's average semantic similarity scores are mainly reduced in magnitude.

Considering the adjusted preferential PageRank similarity, no impact on the semantics can be observed in the Copy-Graph, a nearly constant decrease in the

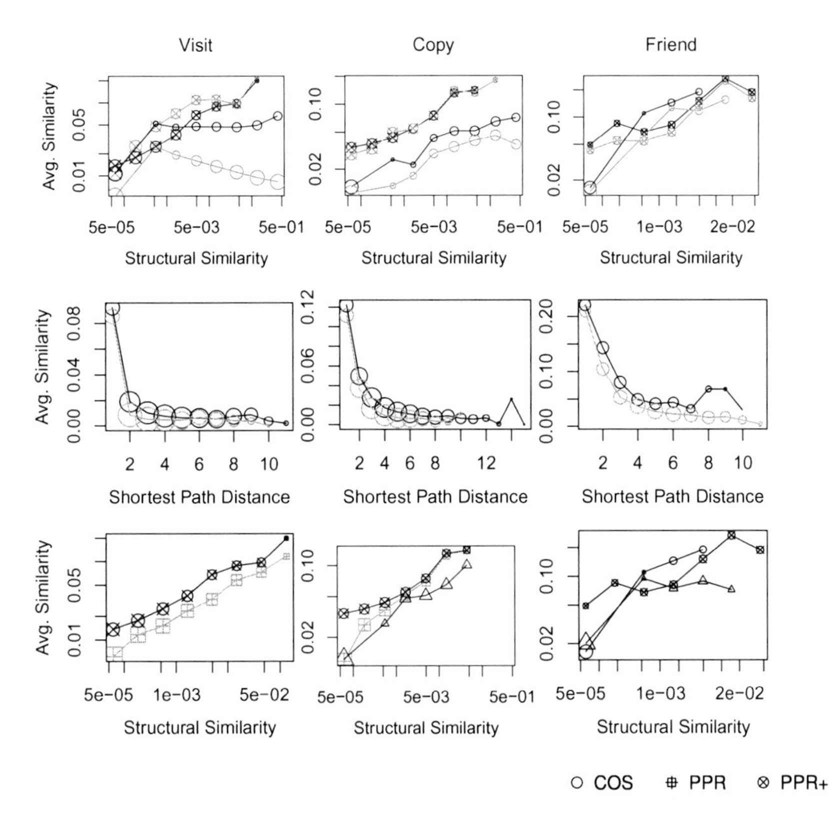

Fig. 5 Average pairwise semantic similarity in BibSonomy, relative to different structural similarity scores (upper row) and relative to the shortest path distance (mid row) in the corresponding directed and undirected networks. The impact considering the preference PageRank relative to the global PageRank is shown in the bottom row. Results for the undirected network variants are depicted in gray.

Friend-Graph, whereas in the Visit-Graph, the corresponding average semantic similarity is mostly increased, loosing monotonicity though.

In Figure 5, the average semantic similarity per shortest path distance in a network is also contrasted to the respective undirected variant. The undirected networks consistently show lowered average semantic similarity per shortest path distance.

*Discussion.*With the preceding analysis, we exemplified, how the proposed experimental set up can be used for assessing the impact of changing certain network parameters, such as the directedness of edges. We conclude, that for the considered networks in BibSonomy the direction of edges significantly contributes to the network semantics and should not be dropped at all.

The Impact of Global PageRank. In Section 4.1, we proposed to adjust the preferential PageRank similarity *PPR* by subtracting the global PageRank component wise.

Using examples from the networks obtained from BibSonomy, we show, that this adjustment significantly increases the corresponding average semantic similarity in Figure 3.

5 Conclusion and Future Work

With the present work, we introduced an experimental framework for assessing the semantics of social networks. The proposed methodology has a broad range of applications, such as *user recommendation* or *community mining* tasks, as it allows semantically grounded pre processing of given networks (e. g., merging different small networks, scaling edge weights, selecting certain groups of users or directedness of networks). The conducted experiments give insights into the semantics of evidence networks from flickr, twitter and BibSonomy and well known similarity metrics. Additionally, the impact of directedness of a network and adjusting the preferential PageRank with the global PageRank is analyzed for the networks obtained from BibSonomy.

Ultimately, the proposed experimental setup allows to formulate the assessment of semantic user relatedness as a regression task, which will be subject to future work.

References

1. Brown, C., Nicosia, V., Scellato, S., Noulas, A., Mascolo, C.: Where Online Friends Meet: Social Communities in Location-based Networks. In: Proc. Sixth International AAAI Conference on Weblogs and Social Media (ICWSM 2012), Dublin, Ireland (2012)
2. Cattuto, C., Benz, D., Hotho, A., Stumme, G.: Semantic Grounding of Tag Relatedness in Social Bookmarking Systems. In: Sheth, A.P., Staab, S., Dean, M., Paolucci, M., Maynard, D., Finin, T., Thirunarayan, K. (eds.) ISWC 2008. LNCS, vol. 5318, pp. 615–631. Springer, Heidelberg (2008)
3. Cohen, T., Widdows, D.: Empirical distributional semantics: methods and biomedical applications. J. Biomed. Inform 42(2), 390–405 (2009)
4. Grefenstette, G.: Finding semantic similarity in raw text: The deese antonyms. In: Fall Symposium Series, Working Notes, Probabilistic Approaches to Natural Language, pp. 61–65 (1992)
5. Hotho, A., Jäschke, R., Schmitz, C., Stumme, G.: Information Retrieval in Folksonomies: Search and Ranking. In: Sure, Y., Domingue, J. (eds.) ESWC 2006. LNCS, vol. 4011, pp. 411–426. Springer, Heidelberg (2006)
6. Islam, A., Inkpen, D.: Second order co-occurrence pmi for determining the semantic similarity of words. In: Proc. of the Int. Conference on Language Resources and Evaluation (LREC 2006), pp. 1033–1038 (2006)
7. Jeh, G., Widom, J.: Simrank: a measure of structural-context similarity. In: Proc. of the Eighth ACM SIGKDD Int. Conference on Knowledge Discovery and Data Mining, KDD 2002, pp. 538–543. ACM, New York (2002)

8. Kaltenbrunner, A., Scellato, S., Volkovich, Y., Laniado, D., Currie, D., Jutemar, E.J., Mascolo, C.: Far From the Eyes, Close on the Web: Impact of Geographic Distance on Online Social Interactions. In: Proc. ACM SIGCOMM Workshop on Online Social Networks (WOSN 2012), Helsinki, Finland (2012)
9. Kulshrestha, J., Kooti, F., Nikravesh, A., Gummadi, K.: Geographic dissection of the twitter network. In: Proc. International AAAI Conference on Weblogs and Social Media (2012)
10. Kwak, H., Lee, C., Park, H., Moon, S.: What is twitter, a social network or a news media? In: Proceedings of the 19th International Conference on World Wide Web, pp. 591–600. ACM (2010)
11. Landauer, T., Dumais, S.: A solution to plato's problem: The latent semantic analysis theory of acquisition, induction, and representation of knowledge. Psychological Review 104(2), 211 (1997)
12. Leicht, E.A., Holme, P., Newman, M.E.J.: Vertex similarity in networks (2005), cite arxiv:physics/0510143
13. Lesk, M.: Word-word associations in document retrieval systems. American Documentation 20(1), 27–38 (1969)
14. Li, N., Chen, G.: Analysis of a Location-Based Social Network. In: Proc. International Conference on Computational Science and Engineering, CSE 2009, pp. 263–270. IEEE Computer Society, Washington, DC (2009)
15. Liben-Nowell, D., Kleinberg, J.: The link-prediction problem for social networks. J. of the American Society for Inf. Science and Technology 58(7), 1019–1031 (2007)
16. Lü, L., Jin, C., Zhou, T.: Similarity index based on local paths for link prediction of complex networks. Physical Review E 80(4), 046122 (2009)
17. Lü, L., Zhou, T.: Link prediction in weighted networks: The role of weak ties. EPL (Europhysics Letters) 89, 18001 (2010)
18. McGee, J., Caverlee, J.A., Cheng, Z.: A Geographic Study of Tie Strength in Social Media. In: Proc. 20th ACM International Conference on Information and Knowledge Management, CIKM 2011, pp. 2333–2336. ACM, New York (2011)
19. Mitzlaff, F., Atzmueller, M., Benz, D., Hotho, A., Stumme, G.: Community Assessment Using Evidence Networks. In: Atzmueller, M., Hotho, A., Strohmaier, M., Chin, A. (eds.) MUSE/MSM 2010. LNCS, vol. 6904, pp. 79–98. Springer, Heidelberg (2011)
20. Mitzlaff, F., Benz, D., Stumme, G., Hotho, A.: Visit Me, Click Me, Be My Friend: An Analysis of Evidence Networks of User Relationships in Bibsonomy. In: Proceedings of the 21st ACM Conference on Hypertext and Hypermedia, Toronto, Canada (2010)
21. Mitzlaff, F., Stumme, G.: Relatedness of given names. Human Journal 1(4), 205–217 (2012)
22. de Sá, H., Prudencio, R.: Supervised link prediction in weighted networks. In: The 2011 Int. Joint Conference on Neural Networks (IJCNN), pp. 2281–2288. IEEE (2011)
23. Sadilek, A., Kautz, H., Bigham, J.P.: Finding Your Friends and Following Them to Where You Are. In: Proc. Fifth ACM International Conference on Web Search and Data Mining, WSDM 2012, pp. 723–732. ACM, New York (2012)
24. Scellato, S., Noulas, A., Lambiotte, R., Mascolo, C.: Socio-spatial properties of online location-based social networks. In: Proceedings of ICWSM 2011, pp. 329–336 (2011)
25. Turney, P.D.: Mining the Web for Synonyms: PMI-IR versus LSA on TOEFL. In: Flach, P.A., De Raedt, L. (eds.) ECML 2001. LNCS (LNAI), vol. 2167, pp. 491–502. Springer, Heidelberg (2001)
26. Weng, J., Lim, E., Jiang, J., He, Q.: Twitterrank: Finding Topic-Sensitive Influential Twitterers. In: Proceedings of the Third ACM International Conference on Web Search and Data Mining, pp. 261–270. ACM (2010)

Social Achievement and Centrality in MathOverflow

Leydi Viviana Montoya, Athen Ma, and Raúl J. Mondragón

Abstract. This paper presents an academic web community, MathOverflow, as a network. Social network analysis is used to examine the interactions among users over a period of two and a half years. We describe relevant aspects associated with its behaviour as a result of the dynamics arisen from users participation and contribution, such as the existence of clusters, rich–club and collaborative properties within the network. We examine, in particular, the relationship between the social achievements obtained by users and node centrality derived from interactions through posting questions, answers and comments. Our study shows that the two aspects have a strong direct correlation; and active participation in the forum seems to be the most effective way to gain social recognition.

1 Introduction

Discussion communities in the format of a Question–Answer (Q&A) site, such as *Yahoo! Answers*[1] have becoming increasingly popular; and the interactions among users and the structure and dynamics of the resultant complex network present a variety of interesting research questions. For example, Rodrigues *et al.* investigated topic management by studying how users choose a topic and their 'tagging' behaviour [1]. Users' participation is of great interest in terms of what motivates users to be active and successful online portals often provide features that enable users to socialise, discuss, chat as well as transfer-share knowledge [2][3]. In addition, Burel *et al.* analysed how these forums operate and attempted to identify the key methods that can lead to assess of quality answers by referring to multiple or specific topics and their complexity [4].

Leydi Viviana Montoya · Athen Ma · Raúl J. Mondragón
Queen Mary University of London, Mile End Road, London E1 4NS
e-mail: {lvmc3,athen.ma,r.j.mondragon}@eecs.qmul.ac.uk

[1] http://uk.answers.yahoo.com

G. Ghoshal et al. (Eds.): *Complex Networks IV*, SCI 476, pp. 27–38.
DOI: 10.1007/978-3-642-36844-8_3

While the benefits of a Q&A forum is evident, particularly when it comes to knowledge exchange and information flow, the reason why people are motivated to help others through these forums is still a bit of a puzzle. For example, MathOverflow[2] is an academic/research community which comprises of members who post high level questions and answers about mathematics. Tausczik *et al.* analysed the relationship between the reputation of users and the quality of their contribution, and examined how the *perceived* reputation might affect the *perceived* quality of the contribution [3]. Tausczik and Pennebaker [5] also studied the motivations for a user to contribute to a Q&A community, they suggested that building reputation is an important incentive for users to participate in the forum. In addition they also noticed that, it is important to find ways to ensure that users contribution are of good quality, and social assessment such as voting seems to be a common way to achieve this.

In this paper, to better understand how users interact with each, we represent the interactions among users in MathOverflow as a network. We find that centrality measures of an individual user give strong indications of social achievement with respect to their social status within the community and quality of their post.

2 MathOverflow

MathOverflow was created by a group of graduates and post-doctoral students from the University of California Berkley in September 2009 [6] and has a total of 22,107 registered users (at the time in which the data was collected in April 2012 [7]). It supports a specialist mathematics Q&A forum, in which posts corresponds to Questions, Answers and Comments. This paper studies all the posts created between September 2010 and April 2012. Each member is provided with 5 social features: *views*, *upvotes*, *downvotes*, *reputation* and *badges*. The *views* attribute describes the number of times the profile of a member and their posts has been viewed through the forum. An *upvote* is a positive score to one's post; and similarly, see fig. 1(a); a *downvote* is a negative score towards one's contribution. *Reputation* is associated with a user's performance, see fig. 1(b). For instance, a user can gain reputation by accumulating *upvotes* and carrying out community and maintenance tasks for the forum, such as editing questions and deleting spam. *Badges* represent the knowledge, expertise, interest or participation of a community member. For example, the users who posted a question with more than 10,000 views are awarded a *Famous question* badge. In addition, badges are divided into Gold, Silver and Bronze to reflect the level of achievement.

2.1 MathOverflow as a Complex Network

This paper studies the interactions among users generated through posting questions, answers and comments. When a question is posted on the forum by a user,

[2] http://MathOverflow.net

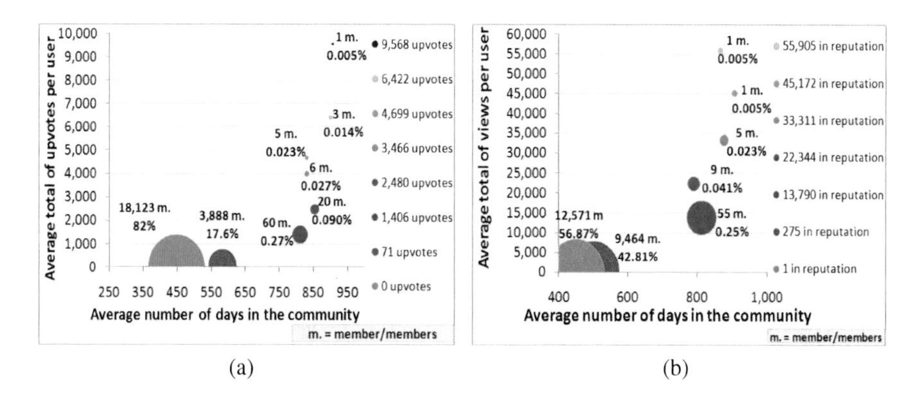

Fig. 1 (a) Upvotes per member between 30-Sept-2009 to 01-April-2012. (b) Reputation score per member between 30-Sept-2009 to 01-April-2012.

other users can answer the question; in addition, they can also comment on a question or an answer. During the period of study, there were a total of 28,360 questions and 78.7% of these questions have at least one answer. The average of answers posted per question was 2.51. The average numbers of questions, answers and comments posted per day is 30.92, 61.09 and 201.45 respectively, describing a significant difference on the volume of comments posted compared to the volume of answers and questions. There are a total of 184,727 comments, 47% were posted on a question and 53% were posted on an answer.

We represent the former interaction as (Q-A) for question-answer interactions and the latter as (QA-C) for question-comment and answer-comment interactions. The interaction among MathOverflow users are described by a graph $\mathscr{G}(\mathscr{E},\mathscr{V})$ which consist of a set of edges $\mathscr{E} = \{n_1,\ldots,n_N\}$ and a set of nodes $\mathscr{V} = \{v_1,\ldots,v_L\}$. The former consists of the users who have taken part in posting and the latter represents a (Q-A) or (QA-C) between two individual users. There are 11,743 nodes and 96,616 unique edges in the MathOverflow network where duplicate edges have been removed.

3 Network Analysis

3.1 Centrality

We first examine the centrality of the MathOverflow network by referring to the *degree*, *betweenness*, *closeness* and *eigenvector* centralities [8].

The degree centrality is based on the number of edges k_i that node n_i has. Nodes with high degree are considered more important in the overall network's structure. The relative importance of the nodes is assessed using the degree distribution. In this network the degree distribution looks like a power law with an exponent $\alpha = -1.2$,

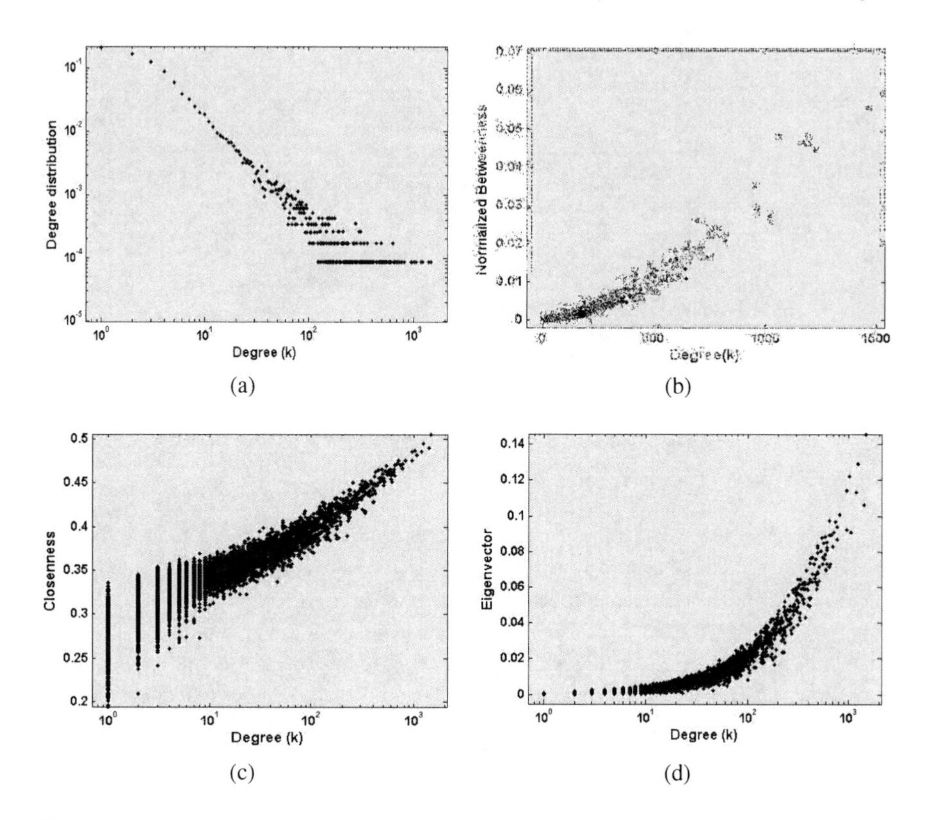

Fig. 2 (a) Degree distribution of nodes in the MathOverflow network. (b) Betweenness of individual nodes against their degree. (c) Closeness of individual nodes against their degree. (d) Eigenvector centrality of individual nodes against their degree.

see Fig. 2(a), which suggest that the network is scale-free. This value is similar to previous studies on co–authorship networks among physicists and computer scientists which is between 0.91 and 1.3 [9]. It has been suggested that if $\alpha < 2$ the network is dominated by few high degree nodes. In the MathOverflow network, most members have less than ten interactions but a small number of users (0.0085%) have interacted with over a thousand members.

The betweenness centrality measures the fraction of geodesic paths that pass through node n_i. In MathOvervlow, the betweenness is low among all the nodes, and the highest (normalised) betweenness is 0.05. We observed a direct correlation between a user's degree and betweenness, as illustrated in Fig. 2(b).

The closeness centrality of a node n_i is the mean geodesic distance from it to every other node. This centrality provides a crude measure on how quickly information spreads. In this network, nodes have a closeness in a range between 0.19 and 0.51; and similar to the betweenness centrality, it is highly related to degree centrality, see fig. 2(c).

The eigenvector centrality measures how well connected a node is and how much direct influence it may have over other well connected nodes in the network. While the degree centrality provides a simple count of the number of connections that node has, the eigenvector centrality assumes that not all connections are equal as the connections to nodes which are themselves influential (with a high degree) will result in a higher score. The eigenvector centrality is the eigenvector corresponding to the largest eigenvalue obtained from the diagonalisation of the adjacency matrix. About 96.5% of the community members have an eigenvector of 0.02 or less, which implies little influence originated from these nodes. However, a small group of 26 users (out of 11,743) have a eigenvector over 0.1, see fig. 2(d).

3.2 Global Network Characteristics

The radius of a network refers to the minimum eccentricity of any node and the diameter is the maximum geodesic distance between two nodes in the network [10]; the MathOverflow network has a radius and a diameter of 4 and 7 respectively. The network density refers to the number of existing edges in the network divided by the total possible number of edges [11], our network has a density of 0.014 which means that there is still a huge potential for the members to interact with other users.

The assortativity of a network is given by

$$r = \frac{\langle kk' \rangle - \langle k \rangle \langle k' \rangle}{\langle k^2 \rangle - \langle k \rangle^2} \tag{1}$$

where the angle brackets $\langle \ldots \rangle$ mean the average over all links and k and k' are the degree of the end nodes of a link. If $r = 0$, there is no correlation between the nodes. If $r > 0$, the network is assortative and disassortative otherwise. The network is disassortative as r has a value of -0.21, and this indicates that high centrality users tend to interact with low centrality users and the presence of a mixing pattern. Network disassortativity is typically found in biological networks and computing and information networks [12].

Transitivity in a social network is associated with the presence of a heightened number of triads (three nodes fully connected) in the network [10]. The transitivity of the network is relatively low with a value of 0.09 so the chance of finding closed triads in the network is quite small.

The Global efficiency of a network is defined as [13]:

$$E_{\text{glob}}(\mathscr{G}) = \frac{1}{N(N-1)} \sum_{i \neq j \in \mathscr{G}} \frac{1}{d_{ij}} \tag{2}$$

where d_{ij} is the geodesic distance from node i to node j and $0 \leqslant E_{glob}(\mathscr{G}) \leqslant 1$. E_{glob} can be seen as a measure of how efficiently information is exchanged over a network, given that all nodes are communicating with all other nodes concurrently [13]. The global efficiency of the network is 0.33.

3.3 Rich–Club

Figure 2 strongly suggest that the nodes of high degree are the most important in the network. To study the interactions between this high degree nodes we use the notion of *rich–club* to examine how the interactions are distributed in this community. High degree nodes that are also highly interconnected to each other are referred as rich nodes [14]. The rich–club coefficient is used to characterise the density of connections between the rich nodes and is given by:

$$\Phi(k) = \frac{2E_{>k}}{N_{>k}(N_{>k}-1)} \tag{3}$$

where $E_{>k}$ corresponds to the number of edges among the $N_{>k}$ nodes having a degree higher than a given value k [15]. $\Phi(k)$ represents the ratio of the real number to the maximally possible number of edges linking the $N_{>k}$ nodes [16].

We considered the top 0.5% of the nodes (58 nodes) with highest degree to define a subgraph of rich nodes, where $k_s = 409$ is the lowest degree in the subgraph. Network measures such as assortativity, clustering coefficient, rich–club coefficient and average shortest path were calculated for the new subgraph.

The subgraph has a high clustering coefficient of 0.8 which means that users who have interacted with a same user tended to make ties (interact) between them. In addition, the subgraph has a negative assortativity value which means that high degree users tend to interact with lower degree users. This could be due to the fact that the rich nodes have a wide range of degrees (between 409 to 1,462). Zhou *et al.* state that the rich–club coefficient $\Phi(k)$ is the ratio of the total actual number of links to the maximum possible number of links between members of the rich-club, and a coefficient of 1 or close to 1 means that the members within the club form a fully connected network [16]. This implies that the top 0.5% of the users with highest degree in the network demonstrated to be highly connected. Nevertheless, Zhang et.al [17] suggest that k_{max}/k_s is a convenient index in complex networks with any degree distribution to show the proportion of links (or degrees) the rich nodes possess in comparison with the rest of nodes in a network. In this paper, k_{max}/k_s = 3.5745 which means rich nodes were far closer connected among them than the majority of the nodes, forming a rich–club.

Zhou *et al.* recommend to compare the clustering coefficient, assortativity and average path length for the rich–club to the same parameters in the network as way to confirm the existence of the rich–club. If these values were quite different then that confirms the rich–club has a different behaviour to that of the whole network [17]. The existence of a rich–club in the network was confirmed as the rich–club has different behaviour than the rest of the network, because its clustering coefficient is much closer to 1 and its geodesic path is small. Both, rich nodes and the whole network have a negative assortativity which is associated to the wide range of existing degree levels inside the network (from 1 to 1,462).

4 Users Attributes and Network Measures

As mentioned previously, MathOverflow has incorporated a range of social features that encourage participation among users. We examine a number of social achievements, namely, reputation, total number of views and number of upvotes; and see how one's social achievements tie in with the centrality measures. Firstly, we define a set of hypotheses with reference to social achievements and centrality. Secondly, the Spearman's rank correlation is used to test the validity of the proposed hypotheses.

Hypothesis 1 – A user's reputation score is closely related to his/her degree centrality. This hypothesis assumes that a user's reputation score is closely related to the number of interactions of the user. On one hand, for a MathOverflow member, reputation is defined as "how much the community trust them?" [18] and a user would need to participate and be active in the network to earn reputation. On the other hand, degree centrality can be seen as a measure of the activity of an actor [19]. As a result, a positive correlation is expected as a user who participates more would have a higher degree and reputation value.

Hypothesis 2 – The total number of views obtained by a user is related to his/her eigenvector centrality. The hypothesis takes into consideration that the measure of eigenvector centrality reflects one's influence in the network and assumes that the higher one's influence, the more likely to attract other users to view your post.

Hypothesis 3 – The number of upvotes obtained by a user is related to his/her closeness centrality. Finally, this refers to one's closeness to other users as an indication of their level of expertise as knowledgeable users are likely to attract followers. Similarly, the number of upvotes obtained by a user shows that such users have been providing reliable or useful posts. A positive correlation is expected as a nodes with a significant volume of upvotes should be more centralised in the network as well as closer to the different kind of users, as the community is comprised of users who would like to learn and share knowledge, it is reasonable to assume that the best voted users are likely to linked to the others.

The Spearman's rank correlation is a statistical measure used to test the direction and strength of the relationship between two variables [20]. The null hypothesis for this kind of test is defined as "there is no relationship between the two sets of data" [21]. The Spearman's rank correlation provides a value r_s which falls between -1 and +1 and a $p-$value to test its statistical significance level [22]. First of all, the data of both variables (x_i and y_i) are ranked separately. Secondly, using the ranked values x_i and y_i, the r_s coefficient is calculated as:

$$r_s = \frac{\sum_i (x_i - \langle x \rangle)(y_i - \langle y \rangle)}{\sqrt{\sum_i (x_i - \langle x \rangle)^2 \sum_i (y_i - \langle y \rangle)^2}} \tag{4}$$

when tied ranks existed on the data [23]. Once r_s is calculated, the existing correlation between the variables can be established depending on which range the absolute value of r_s lies on [20]: *very weak* (0.00 - 0.19); *weak* (0.20 - 0.39); *moderate* (0.40 - 0.59); *strong* (0.60 - 0.79) and *very strong* (0.80 - 1.0). Finally, a significance test is done by evaluating the $p-$value from the student's t distribution, which test the significance of r_s [24]. Table 1 shows the results for the three hypotheses.

Hypothesis 1: Reputation values vs. Degree centrality. The Spearman's rank correlation for the two variables is 0.709. The $p-$value is less than the significance level of the null hypothesis (i.e. $p < 0.05$) which therefore can be rejected. As a result, there is a strong correlation between the reputation score and the degree centrality.

Hypothesis 2: Eigenvector values vs. Total number of views. These variables also showed a strong correlation, and again, the null hypothesis was rejected. This strong correlation means the total number of views is a good indicator of how much the user has been active and participating with popular people in the network. Also, it is important to point out that the eigenvector centrality correlates better with the total number of views than the total number of upvotes, as the latter gives a moderate correlation with $r_s = 0.53$. We can conclude that users with higher eigenvector centrality have more visits to their profile than other users, but they might not have more upvotes than users with lower eigenvector values.

Hypothesis 3: Upvotes vs. Closeness centrality. In contrast, the upvotes and closeness variables are weakly correlated. The null hypothesis that there is no correlation between the variables, was tested at a significance level of 5%. As the p-value is less than the significance level, the null hypothesis was rejected, confirming the weak correlation between Upvotes and Closeness centrality. This result means the users with more popularity in the network for their knowledgable contributions are not necessarily more centralised or close to the general users.

Table 1 Spearman's rank correlation results for the three hypothesis (Significance level of 5%)

Hypothesis	r_s coefficient	Result
1	0.70	Strong correlation
2	0.62	Strong correlation
3	0.28	Weak correlation

5 Behaviours on Subgroups

Specific subgroups in the network were identified based on the tags (topics) the users marked their questions with. When analysing the 3 subgroups with highest number of users (AG–Algebraic Geometry, SQ–Soft Question and NT–Number Theory) they appear to have similar network characteristics as the whole network. These subgroups contain 20%, 19% and 18% of the total users in the network respectively.

The subgroups AG and NT have a diameter larger than the original network, which means the farthest two nodes in AG and NT are more distant; whereas the SQ subgroup has the same diameter as the original network (see Tab. 2).

Table 2 Network parameters for the whole network and the subgroups AG, SQ and NT

Parameter	All network	Subgroup AG	Subgroup SQ	Subgroup NT
Total nodes	11,743	2,310	2,215	2,132
Total unique edges	96,616	12,903	12,392	20,628
Network diameter	7	9	7	8
Network radius	4	5	4	4
Network density	0.0014	0.005	0.005	0.005
Assortativity	-0.2156	-0.1604	-0.1328	-0.1662
Transitivity	0.0955	0.228	0.161	0.235

Identifying clusters in a network was performed in order to identify subgroups with particular behaviour or with nodes which share similar characteristics. The purpose of cluster analysis is to divide the data into groups that are meaningful, useful or both, based only on the information found in the data that describes the objects and their relationships [25]. Given that the network has less than 500,000 nodes the cluster algorithm used is the Clauset-Newman-Moore (CNM) which is based on greedy maximising the modularity function Q [26]. The quality function Q of a network division, defined as

$$Q = \sum_i (e_{ij} - a_i^2) \tag{5}$$

where e_{ij} is the fraction of edges in the network that connect vertices in group i to those in group j, and a_i is the fraction of edges that fall within communities, minus the expected value if edges fall at random [27]. A total of 105 groups were identified when running the algorithm. Most of the clusters are quite small given that 61% of the clusters has just 2 users, see fig. 3(a). The map representing the 105 clusters in the network displayed its 6 biggest groups (denominated as G1, G2, ..., G6) and the small box in the bottom right corner contains 99 clusters with a total of 297 nodes. Also, the 6 clusters have 97.5% of the total users in the MathOverflow network. The subgraphs (clusters) diameter, average geodesic distance and density were calculated for the 6 clusters, in order to characterise them. Cluster G2 has the same diameter than the whole network (7 edges). Clusters G6 and G3 have a diameter of 11. The six clusters show a density lower than 0.006 which is bigger than the density of the whole network (0.0014), although they are significantly low given that the density provides the ratio of direct ties in the network to the total of possible direct ties [28]. The 6 biggest clusters have a small proportion of users with the highest degree values and the highest upvotes counts, with a relevant proportion of users with low degree value and upvotes as followers. The users with a degree higher than 500 or upvotes higher than 750 or with the highest reputations are located only on the 6 biggest clusters (see fig. 3(b)). In contrast, users who have achieved a reputation of 1 (which is given when users input their personal details on the community web page) are located in the biggest clusters as followers of the main users and in the small clusters, Fig. 3(c).

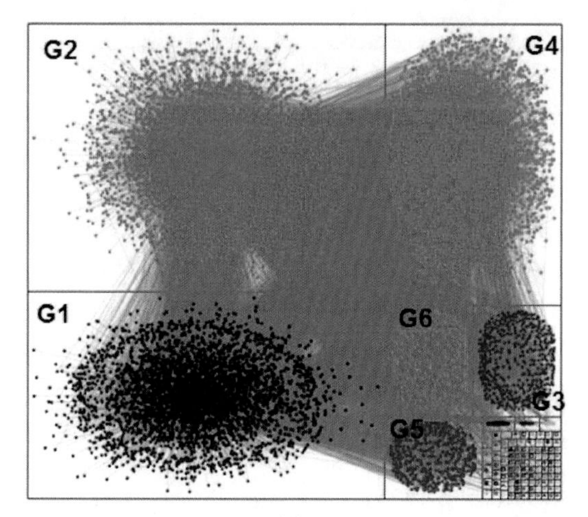

(a)

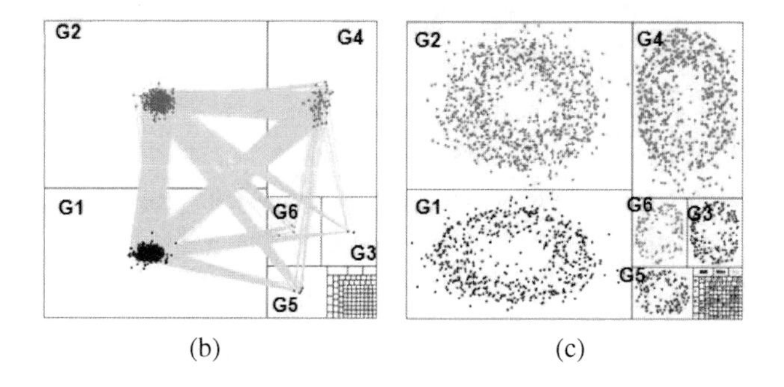

(b) (c)

Fig. 3 (a) Communities derived from the MathOverflow network using modularity and they are mapped using NodeXL [29]. (b) Users with a reputation score of 800 or over and their distribution across the different communities. (c) Users with a reputation score of 1 and their location.

The largest clusters seem to have a similar properties as the whole network. However, when looking at the users attributes in each cluster, there are no specific similitudes between the users in a cluster; the clustering algorithm does not provide clusters which represent a clear criteria of the community division, for example the topic of the questions.

6 Conclusions

Social Network Analysis was used to establish the structure of the academic web community MathOverflow, and to identify main characteristics among its users. The network has a high number of nodes and links and its degree distribution is approximated with a power law, which suggest that the network is scale–free and interactions are dominated by a few high centrality users.

The network is found to be disassortative reflecting that general users are often advised by a *reputable* user who has achieved a certain social status in the network. The network exhibits self-similarity in the sense that when subgroups are defined using individual topics or by clustering using modularity, these subgroups show similar characteristics as the original network.

We study three different hypotheses on the relationship between centrality and social achievements. Spearman's rank correlation on these hypotheses show that a user's reputation, the number of views and the number of upvotes are related to their centrality. It suggests that allowing users to gain social recognition through these features is an effective way to encourage users' participation.

Acknowledgements. The authors would like to thank Ursula Martin, Natasa Milic-Frayling and Eduarda Mendes Rodrigues for their useful comments and ImpactQM for their support.

References

1. Rodrigues, E., Milic-Frayling, N., Fortuna, B.: Social tagging behaviour in community-driven question answering. In: IEEE/WIC/ACM International Conference on Web Intelligence and Intelligent Agent Technology, WI-IAT 2008, vol. 1, pp. 112–119. IEEE (2008)
2. Mendes Rodrigues, E., Milic-Frayling, N.: Socializing or knowledge sharing?: characterizing social intent in community question answering. In: Proceedings of the 18th ACM Conference on Information and Knowledge Management, pp. 1127–1136. ACM (2009)
3. Tausczik, Y., Pennebaker, J.: Predicting the perceived quality of online mathematics contributions from users' reputations. In: Proceedings of the 2011 Annual Conference on Human Factors in Computing Systems, pp. 1885–1888. ACM (2011)
4. Burel, G., He, Y., Alani, H.: Automatic identification of best answers in online enquiry communities. The Semantic Web: Research and Applications, 514–529 (2012)
5. Tausczik, Y., Pennebaker, J.: Participation in an online mathematics community: differentiating motivations to add. In: Proceedings of the ACM 2012 Conference on Computer Supported Cooperative Work, pp. 207–216. ACM (2012)
6. Joel, S.: Cultural anthropology of stack exchange. Hacker News London Meetup - Events (June 2012), http://vimeo.com/37309773
7. MathOverflow: Dumps files (April 1, 2012), http://dumps.mathoverflow.net
8. Newman, M.: The mathematics of networks. Electronic Article (2005), http://www-personal.umich.edu/~mejn/papers/palgrave.pdf
9. Newman, M.: The structure of scientific collaboration networks. Proceedings of the National Academy of Sciences 98(2), 404–409 (2001)

10. Newman, M.: The structure and function of complex networks. SIAM Review 45(2), 167–256 (2003)
11. Faust, K.: Comparing social networks: size, density, and local structure. Metodološki zvezki 3(2), 185–216 (2006)
12. Pastor-Satorras, R., Vázquez, A., Vespignani, A.: Dynamical and correlation properties of the internet. Physical Review Letters 87(25), 258701 (2001)
13. Latora, V., Marchiori, M.: Efficient behavior of small-world networks. Physical Review Letters 87(19), 198701 (2001)
14. Zhou, S., Mondragón, R.J.: The rich-club phenomenon in the internet topology. IEEE Communications Letters 8(3), 180–182 (2004)
15. Colizza, V., Flammini, A., Serrano, M., Vespignani, A.: Detecting rich-club ordering in complex networks. Nature Physics 2(2), 110–115 (2006)
16. Jiang, Z., Zhou, W.: Statistical significance of the rich-club phenomenon in complex networks. New Journal of Physics 10(4), 043002 (2008)
17. Xu, X., Zhang, J., Small, M.: Rich-club connectivity dominates assortativity and transitivity of complex networks. Physical Review E 82(4), 046117 (2010)
18. MathOverflow: Mathoverflow - frequently asked questions. Forum web page (July 2012), http://mathoverflow.net/faq
19. Wasserman, S., Faust, K.: Social network analysis: Methods and applications, vol. 8. Cambridge University Press (1994)
20. Owen, A., Petrie, M., Palipana, A., Green, D., Croft, T., Jones, A., Joiner, S.: Spearman's correlation. Loughborough University (2012), http://www.statstutor.ac.uk/resources/uploaded/spearmans.pdf
21. s.n.: Ib geography notes. Web page (2012), http://www.angelfire.com/ga2/ibgeography/
22. Hossain, L., Wu, A.: Communications network centrality correlates to organisational coordination. International Journal of Project Management 27(8), 795–811 (2009)
23. Lund, A., Lund, M.: Spearman's rank-order correlation. Web page (2012), https://statistics.laerd.com/statistical-guides/spearmans-rank-order-correlation-statistical-guide.php
24. Zar, J.: Significance testing of the spearman rank correlation coefficient. Journal of the American Statistical Association 67(339), 578–580 (1972)
25. Tan, P., Steinbach, M., Kumar, V.: Introduction to data mining. Pearson Addison Wesley (2006)
26. Clauset, A., Newman, M., Moore, C.: Finding community structure in very large networks. Physical Review E 70(6), 066111 (2004)
27. Newman, M.: Fast algorithm for detecting community structure in networks. Physical Review E 69(6), 066133 (2004)
28. Xu, G., Zhang, Y., Li, L.: Web Mining and Social Networking: Techniques and Applications, vol. 6. Springer (2010)
29. Foundation, S.M.R.: Nodexl excel template. Computer Program (2012), http://www.smrfoundation.org/nodexl/

Analysis of Communities Evolution in Dynamic Social Networks

Nikolai Nefedov

Abstract. In this paper we present a framework to study evolution of communities in dynamic networks. A dynamic network is represented by a sequence of static graphs named as network snapshots. We introduce a distance measure between static graphs to study similarity among network snapshots and to detect outlier events. To find a detailed structure within each network snapshot we used a modularity maximization algorithm based on a fast greedy search extended with a random walk approach. Community detection often results in a different number of communities in different network snapshots. To make communities evolution studies feasible we propose a greedy method to match clustering labels assigned to different networks. The suggested framework is applied for analysis of dynamic networks built from real-world mobile datasets.

1 Introduction

The growing spread of smart phones equipped with various sensors makes it possible to record rich-content user data and complement it with on-line processing. Mobile data processing could help people to enrich their social interactions and improve environmental and personal health awareness. At the same time, mobile sensing data could help service providers to understand better human behavior and its dynamics, identify complex patterns of users' mobility, and to develop various service-centric and user-centric mobile applications and services on-demand. One of the first steps in analysis of rich-content mobile datasets is to find an underlying structure of users' interactions and its dynamics by clustering data according to some similarity measures. In cases when data are given in the relational format (causality or dependency relations), e.g., as a network consisting of N nodes and E edges representing some relations among the nodes, then this task may be formulated as a problem of finding

Nikolai Nefedov
ISI Lab., Swiss Federal Institute of Technology, Zurich (ETHZ)
e-mail: nefedov@isi.ee.ethz.ch

G. Ghoshal et al. (Eds.): *Complex Networks IV*, SCI 476, pp. 39–46.
DOI: 10.1007/978-3-642-36844-8_4

communities, i.e., groups of nodes which are interconnected more densely among themselves than with the rest of the network.

The growing interest to the problem of community detection was triggered by the introduction of a new clustering measure called modularity [1]. The direct modularity maximization is known as a NP-hard problem and currently a number of sub-optimal algorithms are proposed, e.g., see [2] and references within. However, most of these methods address static networks partitioning into disjoint communities. On the other hand, in practice communities are dynamic and often overlapping structures. It is especially visible in social networks, where interactions among people and their affiliations to different groups are changing in time.

In this paper we present a framework to study evolution of communities in dynamic networks. A dynamic network is represented by a sequence of static graphs named as network snapshots [3]. We introduce a distance measure between static graphs to study similarity among network snapshots and to detect outlier events. To find a detailed structure within each network snapshot we used a modularity maximization algorithm based on a fast greedy search [4, 5] extended with a random walk approach [6, 7]. Community detection may results in a different number of communities in each network snapshot and in a different labeling of communities within snapshots. To make communities evolution visible we propose a greedy method to match clustering labels assigned to different network snapshots. The paper is organized as follows. In Section 2 we describe a distance measure between networks based on graph Laplacian spectra. A greedy algorithm to match partitions is outlined in Section 3. Analysis of real-world mobile datasets [8] briefly presented in Section 4, followed by conclusions in Section 5.

2 Distance Measure between Networks

To quantify structural properties of dynamic networks a variety of measures has been suggested. For example, in [9] a measure based on Katz-centrality is proposed to analyze time-dependent networks. However, this measure assumes a connected network, which is not always observed in dynamic social or biological networks with a set of disjoint subgraphs. On the other hand, substructure-based measures (e.g, edit-distance, a maximal common subgraph) do not take into account a global structure of a graph. Furthermore, usually only a part of users (nodes) appear in a network snapshot, a total set of nodes is obtained only after the aggregation of all snapshots.

In this paper we use graph spectral methods [10, 11] to characterize global graph structures (e.g., a graph connectivity, disjoint subgraphs) and compare network snapshots defined on a common set of nodes. Graph Laplacian is widely used to describe network structure, but its discrete nature complicates networks comparison. To compare network snapshots aggregated over different time periods we used dynamical systems approach similar to [12, 13].

Let us consider a network of N identical particles (nodes) connected by elastic strings according to an adjacency matrix $\mathbf{A}$ and described by motion equations

$$\ddot{x}_i + \sum_{j=1}^{N-1} A_{ij}(x_i - x_j) = 0, \tag{1}$$

where x_i is the coordinate of the i-th particle. Vibrational frequencies ω_a of this network are defined by eigenvalues $\gamma_a = -\omega_a^2$ of Laplacian $\mathbf{L}_A$ of the matrix $\mathbf{A}$. Laplacian spectrum of a graph is often called a vibrational spectrum [10]. In the following we measure a similarity between two graphs using Laplacian spectra. In particular, we present a spectral density $\rho(\omega)$ of a graph G as a sum of narrow Lorentz distributions [14]

$$\rho(\omega) = K \sum_{a=1}^{N-1} \frac{\gamma}{(\omega - \omega_a)^2 + \gamma^2}, \tag{2}$$

where γ is the width of the Lorentz distributions, K is a normalization coefficient such that $\int \rho(\omega) d\omega = 1$. Using spectral densities (2), a distance $d(G_k, G_m)$ between two graphs G_k and G_m may be defined using the mean square error

$$d_e(G_k, G_m) = \int_0^\infty [\rho_k(\omega) - \rho_m(\omega)]^2 d\omega, \tag{3}$$

or as the inner product of densities

$$d_p(G_k, G_m) = \sum_i \rho_k(\omega_i) \cdot \rho_m(\omega_i). \tag{4}$$

In this paper we use only (4) for networks comparison.

3 Partitions Matching Algorithm

In general, subgraphs matching is a NP-hard problem. In the following we used a greedy matching strategy to find sub-optimal solutions. To match partition labels over all network snapshots we process iteratively two graphs at a time. A simplified description of one iteration is outlined below.

Greedy algorithm to match partition labels in two graphs

Input: partition matrix $P(N,2)$, where $P(:,1)$ is formed by partition labels of a reference graph, $P(:,2)$ consists of community labels of a graph to be matched; labels corresponding to unconnected nodes in $P(:,2)$ are set to zero.

Initialization:
- find indexes of nodes for each of the communities in $P(:,1)$ and $P(:,2)$;
- mark all communities in $P(:,2)$ as unprocessed;

Repeat until all communities are marked as processed in $P(:,2)$:
- select a set of unprocessed communities in $P(:,2)$;

- find a community $c_2(k)$ with the largest number of same labels $l_2^{(m)}$ in $P(:,2)$;
- set $l_2^{(swap)} = l_1(k)$, where $l_1(k)$ corresponds to k-th community label in $P(:,1)$;
- swap labels $l_2^{(m)}$ and $l_2^{(swap)}$ in $P(:,2)$;
- mark community $c_2(k)$ in $P(:,2)$ as processed.

Stop when a maximum number of iterations is reached.

We tested the algorithm using synthesized networks. The greedy matching finds the optimal solution in 80% cases, in other cases solutions are close to the optimal. The complexity of the algorithm is mainly determined by a finite number of selection operations (sorting and swapping), which is in average $\mathcal{O}(N \log N)$.

4 Analysis of Real World Datasets

To analyze mobile users behavior and underlying social structure Nokia/Lausanne organized a mobile data collection campaign (MDCC) at EPFL university campus [15]. Rich-content datasets (including data from mobile sensors, call-logs, users proximity, their locations and etc) are collected from about 200 participants during June/2009-June/2011 [15].

Below we briefly outline applications of the proposed framework for analysis of dynamic social affinity graphs constructed from MDCC voice-call logs. Fig.1 shows network snapshots constructed by aggregating voice-call interactions among MDCC participants during different months. First, we analyzed a similarity among network

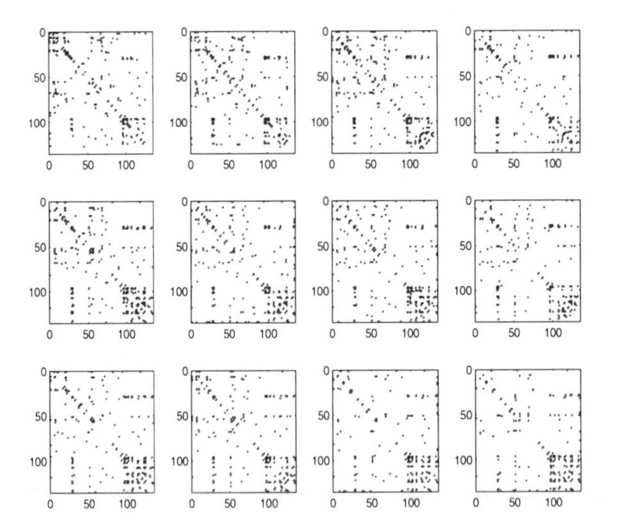

Fig. 1 Dynamics of voice-call activities among MDCC participants during Jan-Dec/2010: adjacency matrices are aggregated over one month period

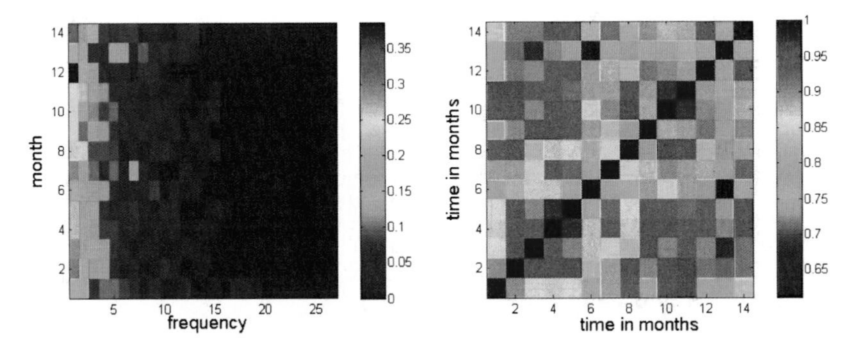

Fig. 2 Normalized vibration spectra (2) of monthly-aggregated network snapshots during Jan/2010-Feb/2011 (left part). Distance d_p among different network snapshots (right part), it clearly shows an outlier at ΔT_{13}.

snapshots using the proposed distance measures described above. As an illustration, vibration spectra (2) and distance (4) among monthly-aggregated network snapshots are depicted at Fig.2, left and right parts, respectively. In particular, at the right part of Fig.2 one can see a high similarity ($d_p = 0.9 \ldots 0.95$) in social interactions for periods $\Delta T_2, \ldots, \Delta T_5$ (Feb-May/2010) and $\Delta T_9, \ldots, \Delta T_{12}$ (Sept-Dec/2010). Since a significant part of MDCC participants are students, these similar behavior patterns most probably correspond to session periods at the EPFL university. Also, one can clearly see an abnormality in social interaction at ΔT_{13} (Jan/2011). Detailed inspection of the data revealed that during this period most of the participants were contacted by one of the organizers about the MDCC updated conditions. For the following analysis we removed network connections relevant to detected outlier events.

To find communities in each network snapshot we used the algorithm [4] extended with a random walk [6, 7]. Communities detected in the voice-call network for data aggregated over the whole data collection campaign are shown by different colors at Fig.3.

Next, we applied the community detection algorithm for network snapshots built from monthly data. Connectivity among participants and their numbers are different at each snapshot, it results in a different number of communities shown by different colors at each snapshot (Fig.4). Furthermore, even in cases when some nodes (users) happen to belong to the same community at different time periods, their community labels assigned by community detection at different network snapshots may not necessarily coincide. As an illustration, color-coded community labels in different network snapshots corresponding to different months are shown at Fig.5, left. Dark blue color here (marked by zero at color bar) indicates no-calls intervals for a user within the participants set.

Hence, to analyze a community evolution we need to find a set of clustering labels at each network snapshot which gives the best match to clustering labels for a reference case. As a reference for MDCC datasets we used snapshots ag-

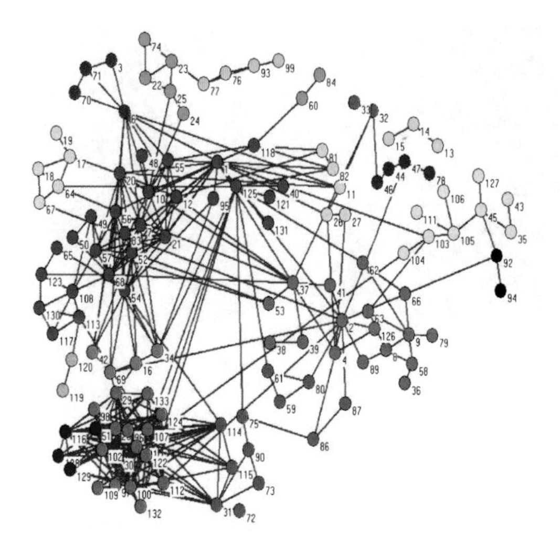

Fig. 3 Communities detection in voice-call network; data are aggregated over the whole MDCC period

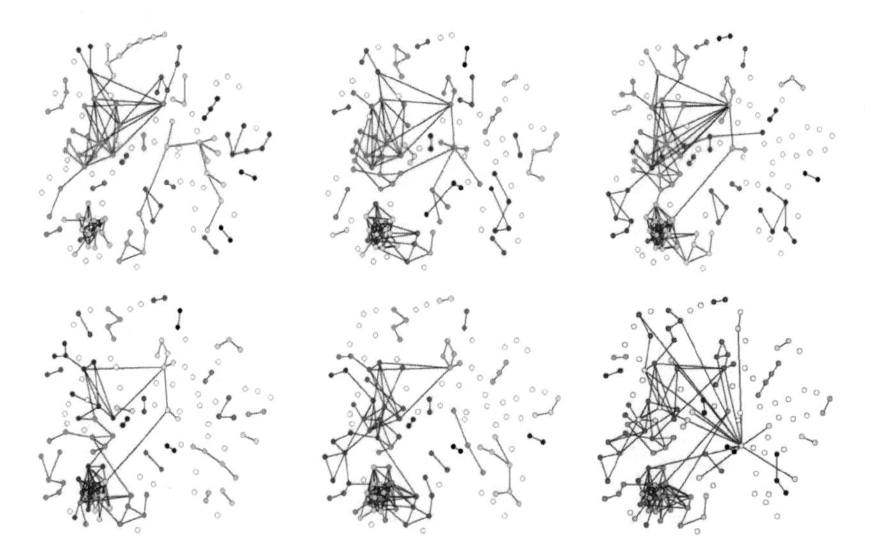

Fig. 4 Communities detection in voice-call networks, data are aggregated over one month period. Upper row: Jan-March/2010; lower row: Apr-June/2010.

gregated over the whole period (cf. Fig.3). Fig.5 (right part) depicts communities within monthly snapshots with community labels matched to the reference network. Columns on the left from color bars at Fig.5 present communities detected in the reference network. All participants are re-ordered according to community labels derived from the reference network. As one can see, after re-labeling the evolution of

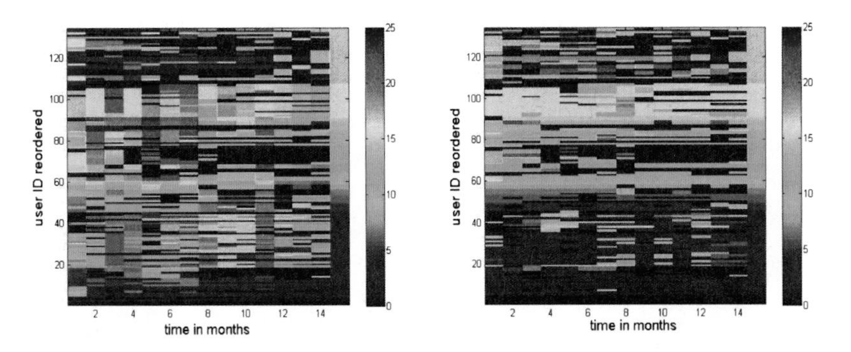

Fig. 5 Evolution of communities in MDCC voice-call social network during 14 months: color-coded community labels for 134 users in monthly-aggregated snapshots before (left part) and after (right part) re-labeling. Dark blue color (marked by zero at color bar) indicates no-calls intervals. The 15th column on both figures presents the assignment of community labels in the aggregated over the whole period network.

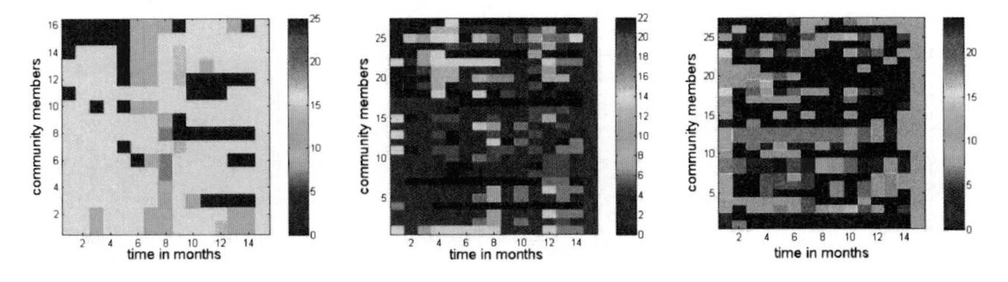

Fig. 6 Evolution of voice-call activity of users within their own communities in time. Examples of communities with dominating intra- and inter-community activities are shown on the left and on the right, respectively. Color bar represents community labels.

communities in time became clearly visible. Examples of communities with intra- and inter-communities interactions are depicted at Fig.6. One of the observations here is that communities detected in networks built from all aggregated data may be misleading. For example, the community shown at Fig.6 (right) actually is not observable in monthly network snapshots. In fact, it hardly may be called a homogenous community due to prevailing inter-community interactions, while users interactions within this community are sparse and not stable. It looks that this community actually is an artifact appeared due to data aggregation over a long period. On the other hand, the community at Fig.6 (left) reveals the stable structure at both, monthly and over the year, time scales.

5 Conclusions

In this paper we introduced a distance measure between network snapshots and applied it to study dynamics of communities in real-world mobile datasets. The proposed method allowed us to find outliers and clean the data. Community detection results in a different number of communities at different network snapshots. To match clustering labels at different snapshots we proposed a suboptimal greedy re-labeling method, verified it on synthesized networks and then applied it for real-world mobile data. The proposed method allowed us to remove artifacts in community detection due to data aggregation.

References

1. Newman, M.E.J., Girvan, M.: Finding and evaluating community structure in networks. Physical Review E 69, 026113 (2004)
2. Fortunato, S.: Community detection in graphs. Physics Reports 486, 75–174 (2011)
3. Spiliopoulou, M.: Evolution in Social Networks: A Survey. In: Social Network Data Analytics, pp. 149–175. Springer Science+Business Media, LLC (2011)
4. Newman, M.E.J.: Fast algorithm for detecting community structure in networks. Physical Review E 69, 066133 (2004)
5. Blondel, V., Guillaume, J.L., Lambiotte, R., Lefebvre, E.: Fast unfolding of communities in large networks. Journal of Statistical Mechanics: Theory and Experiment 1742-5468(10), P10008+12 (2008)
6. Lambiotte, R., Delvenne, J.C., Barahona, M.: Laplacian Dynamics and Multiscale Modular Structure in Networks, ArXiv:0812.1770v3 (2009)
7. Nefedov, N.: Multiple-Membership Communities Detection and its Applications for Mobile Networks. In: Applications of Digital Signal Processing, pp. 51–76. InTech (November 2011)
8. Nokia Mobile Data Challenge Campaign, `http://research.nokia.com/page/12000`
9. Grindrod, P., Higham, D.J., Parsons, M.C., Estrada, E.: Communicability across evolving networks. Phys. Rev. E 83, 046120 (2011)
10. Chung, F.R.K.: Spectral Graph Theory. CMBS Lectures Notes 92. AMS (1997)
11. Fay, D., et al.: Weighted Spectral Distributions: A Metric for structural Analysis of Networks. In: Statistical and Machine Learning Approaches for Network Analysis, pp. 153–190. John Wiley & Sons Inc., NY (2012)
12. Arenas, A., Diaz-Guilera, A., Kurths, J., Moreno, Y., Zhou, C.: Synchronization in complex networks. Physics Reports 469, 93–153 (2008)
13. Nefedov, N.: Applications of System Dynamics for Communities Detection in Complex Networks. In: IEEE Int. Conf. on Nonlinear Dynamics and Sync. (INDS 2011) (2011)
14. Ipsen, M., Mikhailov, A.: Evolutionary reconstruction of networks. Physical Review E 66, 046109 (2002)
15. Kiukkonen, N., Blom, J., Dousse, O., Gatica-Perez, D., Laurila, J.: Towards Rich Mobile Phone Datasets: Lausanne Data Collection Campaign. In: Proc. ACM Int. Conf. Pervasive Services, Berlin (2010)

Entropy Production in Stationary Social Networks

Haye Hinrichsen, Tobias Hoßfeld, Matthias Hirth, and Phuoc Tran-Gia

Abstract. Completing their initial phase of rapid growth social networks are expected to reach a plateau from where on they are in a statistically stationary state. Such stationary conditions may have different dynamical properties. For example, if each message in a network is followed by a reply in opposite direction, the dynamics is locally balanced. Otherwise, if messages are ignored or forwarded to a different user, one may reach a stationary state with a directed flow of information. To distinguish between the two situations, we propose a quantity called *entropy production* that was introduced in statistical physics as a measure for non-vanishing probability currents in nonequilibrium stationary states. The proposed quantity closes a gap for characterizing social networks. As major contribution, we present a general scheme that allows one to measure the entropy production in arbitrary social networks in which individuals are interacting with each other, e.g. by exchanging messages. The scheme is then applied for a specific example of the R mailing list.

1 Introduction

Due to the rapid growth of social media in the last decade, many theoretical studies have been focused on the growth dynamics of social networks [1]. In such a social network, individuals are connected to each other (e.g. friends in facebook, sender and receiver in mail networks) and there is an interaction between the individuals (e.g. exchanging messages). Hence, beside the network topology, the interaction

Haye Hinrichsen
University of Würzburg, Department of Physics and Astronomy, Am Hubland, 97074 Würzburg, Germany
e-mail: hinrichsen@physik.uni-wuerzburg.de

Tobias Hoßfeld · Matthias Hirth · Phuoc Tran-Gia
University of Würzburg, Institute of Computer Science, Chair of Communication Networks, Am Hubland, 97074 Würzburg, Germany
e-mail: tobias.hossfeld@uni-wuerzburg.de

G. Ghoshal et al. (Eds.): *Complex Networks IV*, SCI 476, pp. 47–58.
DOI: 10.1007/978-3-642-36844-8_5

among invididuals characterize the social network. However, recent observations [2] indicate that many social networks are approaching a plateau of constant size, e.g. due to logistic growth models and resulting upper population bounds like in [3]. In such a matured state, the dynamics of the network are approximately stationary in a statistical sense, meaning that the network topology as well as the probability for receiving and sending messages do not change in the long-term limit.

The dynamics of a stationary state of a network is not uniquely given, rather there is a large variety of possible realizations. For example, the three individuals shown in Fig. 1 may send messages (a) randomly in both directions or (b) in clockwise direction. In both situations the individuals send and receive messages at constant rate, meaning that the network is statistically stationary. However, in the first case the dynamics is locally balanced between pairs of users, while in the second case there is a directed current of messages flowing clockwise through the system.

In the present work we introduce a new type of quantity, called *entropy production* in statistical physics, to characterize the stationary properties of arbitrary social networks. To this end, we associate with each pair of individuals i, j a quantity H_{ij} called entropy, which depends on the number of messages sent from i to j and vice versa. The entropy H_{ij} measures the directionality of the information exchange and vanishes for perfectly balanced communication. Defining the entropy production of a node as the sum over the entropy of all its links, one can identify nodes contributing preferentially to balanced or unidirectional information transfer.

The concept of entropy production requires to make certain assumptions about the dynamics of the network. In particular, we ignore possible correlations between the messages by assuming that the individuals communicate randomly at constant rates. With this assumption each message sent from node i to node j increases the entropy by [4, 5, 6, 7]

$$\Delta H_{ij} = \ln w_{ij} - \ln w_{ji}, \tag{1}$$

where w_{ij} and w_{ji} are the rates for messages from i to j and in opposite direction, respectively. In physics, this quantity can be interpreted as the minimal entropy produced by a machine that keeps the network running. In computer science this interpretation is irrelevant since realistic networks produce much more entropy in the environment. However, as we will demonstrate in the present work, the entropy production is a useful measure to characterize the stationary properties of the network as, for example, to distinguish the situations (a) and (b) in Fig. 1.

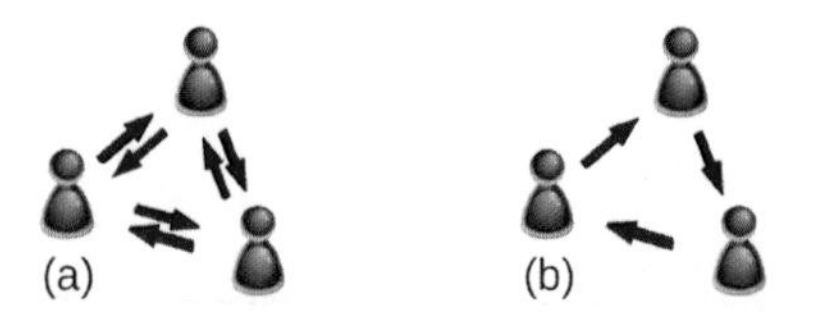

Fig. 1 Example of a stationary network with three users. (a) Each individual sends messages to randomly selected neighbors, leading to a statistically balanced stationary state with vanishing entropy production. (b) The individuals send messages to only one neighbor, generating a stationary but directed flow of information with positive entropy production.

Equation (1) is trivial to evaluate if the rates w_{ij} and w_{ji} are known. However, in realistic networks with data taking over a finite time span T, only the number of messages n_{ij} and n_{ji} exchanged between pairs of nodes are known. Although it is tempting to replace the rates w_{ij} by the relative frequencies n_{ij}/T and to approximate the entropy production by $\Delta H_{ij} = \ln n_{ij} - \ln n_{ji}$, it is easy to see that this approximation would diverge as soon as one of the count numbers vanishes. Therefore, the paper deals to a large extent with the question how we can reasonably reconstruct the rates from the given number of messages.

The remainder of this paper is structured as follows. Section 2 introduces variables describing the observed data from a measurement campaign of a social network. Further, assumptions on the network dynamics are summarized. Based on that, Section 3 defines the entropy production which is based on the (unknown) message rate between any two individuals of the social network. An estimator of the rates based on the observed measurement data is introduced by means of Bayesian inference. Section 4 presents appropriate choice of the prior distribution for small-world networks, in which the number of messages follows a power-law distribution. Relevant parameters of the prior distribution are calculated which finally allows computing the entropy production. The general scheme is summarized for social networks manifesting small-world characteristics on the number of messages. In Section 5, the general scheme is applied exemplarily to the R mailing list. Section 6 revisits related work in order to show that entropy production fills a gap in characterizing social networks. Finally, Section 7 concludes the work and gives an outlook on next steps in this research direction.

2 Stationary Network Dynamics

2.1 *Observed Data*

Let us consider a social network of individuals communicating by directed messages (e.g. emails, Twitter or Facebook messages). Suppose that we monitor M messages over a finite time span, recording sender and receiver id's in a file. Such a data set can be represented as a graph of nodes (individuals) connected by directed links (messages) as depicted in Fig. 2. Enumerating the individuals in the list by $i = 1 \ldots N$, let n_{ij} be the number of messages sent from i to j. These numbers constitute a $N \times N$ connectivity matrix which is the starting point for the subsequent analysis. If at least one message is sent from i to j, the two nodes are said to be connected by a directed link. Obviously, the number of recorded messages M and the total number of directed links L are the following using the Kronecker delta $\delta_{a,b}$.

$$M = \sum_{i,j=1}^{N} n_{ij}, \qquad L = \sum_{i,j=1}^{N} (1 - \delta_{0,n_{ij}}) \tag{2}$$

Note that $M \geq L$ since two individuals can communicate several times during the observation period. The statistics of multiple communication is described by the probability distribution

$$P(n) := \frac{\sum_{i,j=1}^{N} \delta_{n,n_{ij}}}{N(N-1)} \tag{3}$$

of the matrix elements n_{ij}. In the present work we are particularly interested in social media networks with a small-world topology, where this distribution follows a power law,

$$P(n) \sim n^{-1-\alpha}. \tag{4}$$

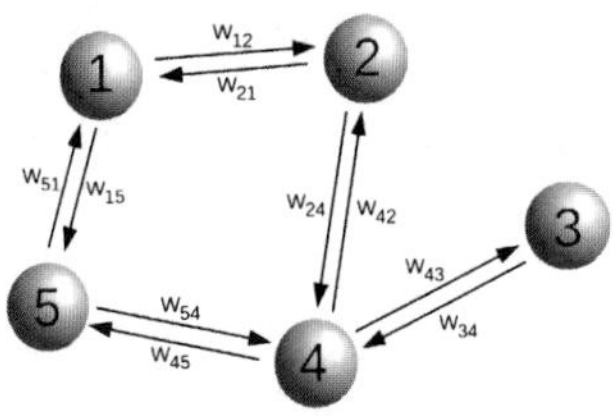

$$\{w_{ij}\} = \begin{pmatrix} 0 & 0.3 & 0 & 0 & 2.0 \\ 0.5 & 0 & 0 & 0.5 & 0 \\ 0 & 0 & 0 & 3.0 & 0 \\ 0 & 0.7 & 0.2 & 0 & 1.5 \\ 0.2 & 0 & 0 & 1.5 & 0 \end{pmatrix}, \{n_{ij}\} = \begin{pmatrix} 0 & 0 & 0 & 0 & 5 \\ 1 & 0 & 0 & 1 & 0 \\ 0 & 0 & 0 & 4 & 0 \\ 0 & 1 & 1 & 0 & 3 \\ 0 & 0 & 0 & 2 & 0 \end{pmatrix}$$

Fig. 2 Example of a directed social network with $N = 5$ participants. It is assumed that node i sends messages to node j randomly with the rate w_{ij}. Observing the network for a finite time span the number of recorded messages from i to j is n_{ij}. In order to compute the entropy production, one has to estimate the unknown rates w_{ij} from the numbers n_{ij}.

2.2 *Assumptions on Network Dynamics*

In a realistic social network the messages are causally connected and mutually correlated. As this information is usually not available and requires semantic analysis of the messages, let us consider the messages as uncorrelated instantaneous events which occur randomly like the clicks of a Geiger counter. More specifically, we start with the following assumptions:

- *Stationarity:* We assume that the size of the social network is approximately constant during data taking. This means that the total number N_{tot} of participants in the system is constant. This assumption is e.g. valid for networks following a logistic growth model [3]. Note that N_{tot} may be larger than the actual number of participants N communicating during data taking.
- *Effective rates:* Messages are sent from node i to node j at a constant *rate* (probability per unit time), denoted as $w_{ij} \geq 0$.
- *Reversibility:* If node i is can communicate with note j, node j can also communicate with node i. This assumption is typically true in social networks. Hence if w_{ij} is nonzero, then the rate in opposite direction w_{ji} is also nonzero.

With these assumptions, the average number of communications from i to j is given by $\langle n_{ij} \rangle = w_{ij}T$, where T denotes the observation time.

3 Entropy Production

3.1 *Definition*

As outlined above, each message sent from node i to j produces an entropy of

$$\Delta H_{ij} := \ln \frac{w_{ij}}{w_{ji}} . \tag{5}$$

Since node i sends n_{ij} messages to node j during the observation period, the total entropy produced by messages $i \to j$ is given by $n_{ij} \Delta H_{ij}$, while messages in opposite direction produce the entropy $n_{ji} \Delta H_{ji}$. Adding the two contributions we obtain the link entropy

$$H_{ij} = n_{ij} \Delta H_{ij} + n_{ji} \Delta H_{ji} = (n_{ij} - n_{ji}) \ln \frac{w_{ij}}{w_{ji}} . \tag{6}$$

This entropy is symmetric ($H_{ij} = H_{ji}$) and can equally be attributed to the corresponding nodes, allowing us to define an entropy production per node

$$H_i = \frac{1}{2} \sum_{j=1}^{N} H_{ij} \tag{7}$$

as well as the entropy production of the total network

$$H = \sum_i H_i = \frac{1}{2} \sum_{i,j=1}^{N} H_{ij} . \tag{8}$$

3.2 *Naïve Estimate*

The entropy production depends on the message numbers n_{ij} and the rates w_{ij}. While the numbers n_{ij} can be determined directly from the given data, the rates w_{ij} are usually not known in advance. Of course, in the limit of infinite observation time the relative frequencies of messages converge to the corresponding rates, i.e.

$$w_{ij} = \lim_{T \to \infty} \frac{n_{ij}}{T} . \tag{9}$$

For finite observation time the count numbers n_{ij} are scattered around their mean value $\langle n_{ij} \rangle = T w_{ij}$. Therefore it is tempting to approximate the entropy production by replacing the ratio of the rates with the ratio of the relative frequencies, i.e.

$$H_{ij}^{\text{naive}} \approx (n_{ij} - n_{ji}) \ln \frac{n_{ij}}{n_{ji}} . \tag{10}$$

However, this naïve estimator is useless for two reasons. Firstly, the nonlinear logarithm does not commute with the linear average and is thus expected to generate systematic deviations. Secondly, in realistic data sets there may be one-way communications with $n_{ij} > 0$ and $n_{ji} = 0$, producing diverging contributions in the naïve estimator (10). However, observing no messages in opposite direction does not mean that the actual rate is zero, it only means that the rate is small. In the following we suggest a possible solution to this problem by using standard methods of Bayesian inference, following similar ideas recently addressed in a different context [8].

3.3 Bayesian Inference

As the messages are assumed to occur randomly and independently like the clicks of a Geiger counter, we expect the number of messages n for a given rate w to be distributed according to the Poisson distribution

$$P(n|w) = \frac{(Tw)^n e^{-Tw}}{n!}, \tag{11}$$

where T is the observation time. But instead of n for given w, we need an estimate of the rate w for given n. According to Bayes formula [9] the corresponding conditional probability distribution is given by the posterior

$$P(w|n) = \frac{P(n|w)P(w)}{P(n)}, \tag{12}$$

where $P(w)$ is the prior distribution and

$$P(n) = \int_0^\infty dw\, P(n|w)P(w) \tag{13}$$

is the normalizing marginal likelihood. The prior distribution expresses our belief how the rates are statistically distributed and introduces an element of ambiguity as will be discussed below. Having chosen an appropriate prior the expectation value $\langle \ln w \rangle$ for given n reads

$$\langle \ln w \rangle_n = \int_0^\infty dw\, \ln w P(w|n). \tag{14}$$

This allows us to estimate the entropy production of the directed link $i \to j$ by

$$H_{ij} \approx (n_{ij} - n_{ji}) \Big[\langle \ln w \rangle_{n_{ij}} - \langle \ln w \rangle_{n_{ji}} \Big]. \tag{15}$$

As we will see, this estimator does not diverge if $n_{ji} = 0$.

4 Small-World Networks

4.1 Choice of the Prior Distribution

The prior should be as much as possible in accordance with the available data. In the example to be discussed below, where we investigate a small-world network with message numbers distributed according to Eq. (4) with an exponent $\alpha > 1$, it would be natural to postulate a power-law distribution of the rates $P(w) \sim w^{-1-\alpha}$. Since such a distribution can only be normalized with a suitable lower cutoff, a natural choice for the prior would be the inverse gamma distribution

$$P(w) = \frac{\beta^{\alpha} w^{-\alpha-1} e^{-\beta/w}}{\Gamma(\alpha)}, \tag{16}$$

where the parameter β plays the role of a lower cutoff for the rate w. With this prior distribution the integration can be carried out, giving the posterior

$$P(w|n) = \frac{(\beta/T)^{\frac{\alpha-n}{2}} w^{n-\alpha-1} e^{-Tw-\frac{\beta}{w}}}{2K_{n-\alpha}(z)}, \tag{17}$$

where $K_{\nu}(z)$ is the modified Bessel function of the second kind and $z = 2\sqrt{\beta T}$. Inserting this result into Eq. (14) we obtain an estimate of $\ln w$ for given n, namely

$$\langle \ln w \rangle_n = \frac{1}{2} \ln \frac{\beta}{T} + \frac{K_{n-\alpha}^{(1,0)}(z)}{K_{n-\alpha}(z)}, \tag{18}$$

where $K_{\nu}^{(1,0)}(z) = \frac{\partial}{\partial \nu} K_{\nu}(z)$. The estimator for the entropy production is given by

$$H_{ij} \approx (n_{ij} - n_{ji}) \left[\frac{K_{n_{ij}-\alpha}^{(1,0)}(z)}{K_{n_{ij}-\alpha}(z)} - \frac{K_{n_{ji}-\alpha}^{(1,0)}(z)}{K_{n_{ji}-\alpha}(z)} \right]. \tag{19}$$

4.2 Estimating the Cutoff Parameter $z = 2\sqrt{\beta T}$

Eq. (18) depends on the exponent α, which can be obtained from the distribution of messages per link, and the lower cutoff β, which can be determined as follows. On the one hand, the probability to have no link between two nodes for a given rate w is $1 - P(0|w)$. Therefore, the total number of links L can be estimated by

$$L \approx L_{\text{tot}} \int_0^{\infty} dw \Big(1 - P(0|w)\Big) P(w) = L_{\text{tot}} \Big(1 - \frac{2(T\beta)^{\alpha/2}}{\Gamma(\alpha)} K_{\alpha}(2\sqrt{T\beta})\Big). \tag{20}$$

Here $L_{tot} = N_{\text{tot}}(N_{\text{tot}} - 1)$ is the unknown total number of potential links in the stationary network which may exceed the actual number of links L established during the finite observation time T. On the other hand, it is obvious that the total number of messages M can be estimated by

$$M \approx L_{\text{tot}} \sum_{n=0}^{\infty} n \int_0^{\infty} dw\, P(n|w)P(w) = L_{\text{tot}} \frac{T\beta}{\alpha - 1}. \tag{21}$$

This relation can be used to eliminate L_{tot}, turning Eq. (20) into

$$\frac{Lz^2}{4M(\alpha-1)} \approx 1 - \frac{2(z/2)^{\alpha}}{\Gamma(\alpha)} K_{\alpha}(z). \tag{22}$$

For given M, L, α this approximation interpreted as an equation allows us to numerically determine $z = 2\sqrt{\beta T}$.

4.3 *Summary of the Procedure*

The procedure to calculate the entropy production can be summarized as follows:

1. In the given data set of M messages, identify all participants (nodes) and label them from $1,\ldots,N$.
2. Determine the numbers n_{ij} how often a message is sent from i to j and count the number L of nonzero entries (links) in the matrix $\{n_{ij}\}$.
3. Plot a histogram of the numbers n_{ij}. If it exhibits a power law $P(n) \sim n^{-1-\alpha}$ estimate the exponent α.
4. Solve Eq.(22) numerically for z.
5. Compute the numbers $\chi_n = K^{(1,0)}_{n-\alpha}(z)/K_{n-\alpha}(z)$.
6. Associate with each directed link $i \to j$ the entropy production $H_{ij} = (n_{ij} - n_{ji})(\chi_{n_{ij}} - \chi_{n_{ji}})$.
7. Compute H_i and H according to Eqs. (7) and (8).

5 Example: Mailing List Archive

To demonstrate the concepts introduced above, we analyzed the mailing lists archive for the programming language **R** [10], recording senders and receivers of all messages over the past 15 years. In this mailing list $N = 23462$ individuals (nodes) have exchanged $M = 168778$ directed comments (undirected activities like opening a new thread are ignored). The connectivity matrix n_{ij} has $L = 114713$ nonzero entries (links). Their statistical distribution shown in Fig. 3 confirms a small-world topology with $\alpha \approx 2$. Interestingly, the node degree distribution of outgoing and incoming links in Fig. 3(b) seems to exhibit slightly different exponents. A similar phenomenon was observed some time ago in email communication networks [11].

Since H_{ij} depends on two integers n_{ij} and n_{ji}, the entropy production of a link produces a discrete set of values. The upper panel of Fig. 4 shows how these values are distributed and how often they occur. As can be seen, the entropy production varies over five orders of magnitude and is distributed irregularly with count numbers ranging from 1 to 10^4.

Let us now turn to the question how the node entropy $H_i = \frac{1}{2}\sum_{j=1}^{N} H_{ij}$ is correlated with the number of outgoing and incoming messages $n_i^{\text{out}} = \sum_j n_{ij}$, $n_i^{\text{in}} = \sum_j n_{ji}$. Since the entropy production is expected to grow with the number of messages, it is reasonable to define the node entropy production per message

$$h_i := \frac{H_i}{n_i^{\text{out}} + n_i^{\text{in}}} . \tag{23}$$

Figure 5 shows how the entropy production per node is distributed depending on the number of sent and received messages. As expected, the entropy is minimal if these numbers coincide. Plotting the entropy production of a node versus the difference of outgoing and incoming messages $\Delta n_i = n_i^{\text{out}} - n_i^{\text{in}}$ one finds again an asymmetric distribution (see right panel). This indicates that nodes with a large number of outgoing links tend to produce less entropy per message than individuals who preferentially receive messages.

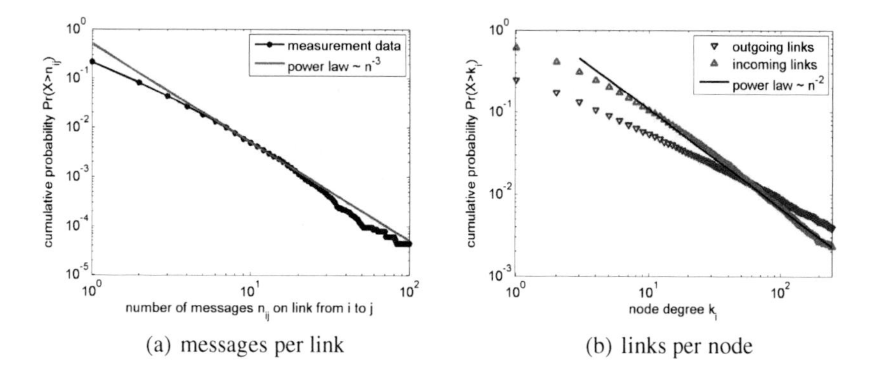

(a) messages per link (b) links per node

Fig. 3 (a) Probability $Pr(X > n_{ij})$ that a directed link $i \to j$ carries more than n_{ij} messages. The data is consistent with a power law $P(n_{ij}) \sim n_{ij}^{-3}$. (b) Probability $Pr(X > k_i)$ that a node i is connected with more than k_i outgoing or incoming links. For incoming links, it is $P(k_i) \sim k_i^{-2}$.

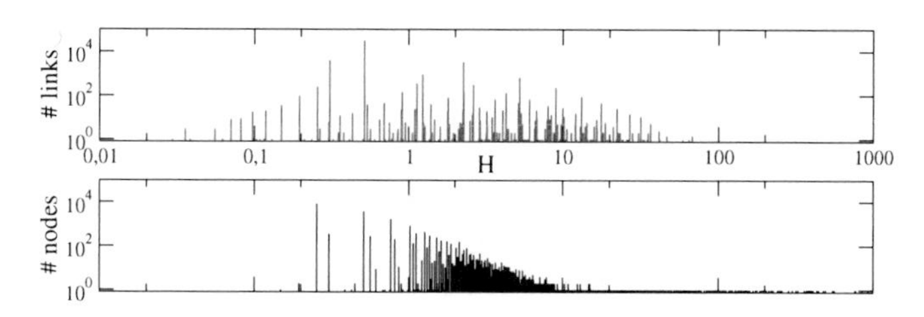

Fig. 4 Upper panel: Histogram of link entropy H_{ij}. Lower panel: Histogram of node entropy H_i.

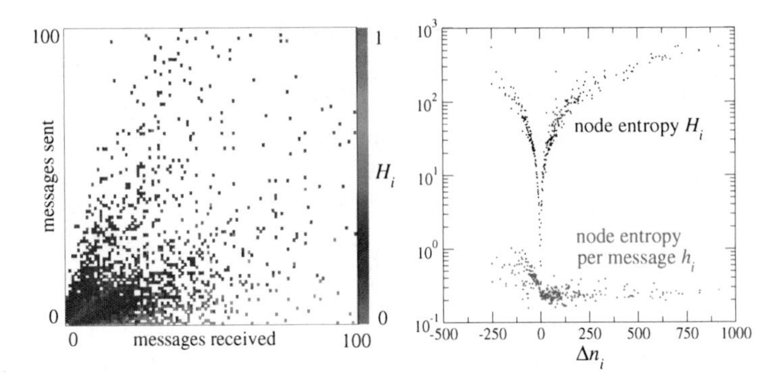

Fig. 5 Left: Average entropy production per node represented by a continuous color scale in a two-dimensional plane spanned by the number of incoming and outgoing messages. Right: Average entropy production of a node (black) and the same data divided by the total number of incoming and outgoing messages (red) as a function of the difference between outgoing and incoming messages.

6 Discussions on Related Work

In literature, various measures of the characteristics of complex networks exist. We briefly revisit them in order to show that entropy production fills a gap for measuring the directionality of the information exchange and quantifying the balance of communication or interaction. The quantities introduced in literature analyze mainly the topology of the graph itself by means of the adjacency matrix $\mathscr{A}$ with elements $\mathscr{A}_{ij} = 1 - \delta_{0,n_{ij}}$. Extensions of several quantities exist for weighted networks. In that case, the directed link connecting the nodes i and j are weighted by the message rate w_{ij}. Instead of $\mathscr{A}$, the message rate matrix $\mathscr{W}$ is used. Thereby, beyond the topological effects, the metrics which allow to work on weighted networks give insights into the structure of the message diffusion, too. Those metrics are to be analyzed with the network entropy H for different network topologies and message exchange models $\mathscr{W}$. Future concerns an analysis of those metrics with the entropy production for different network topologies and message exchange models $\mathscr{W}$.

Principal graph characteristics. The basic quantities are the in- and out-degree of nodes corresponding to the number of incoming and outgoing links of nodes. We observe a strong correlation of the node degree with entropy production for the example of the R mailing list. However, a closer look in the previous section revealed that nodes with a large number of outgoing links tend to produce less entropy per message. Future work investigates for which kind of network topologies and message exchange models those quantities are correlated. Other principal characteristics of nodes are eccentricity and local clustering coefficient. Global network metrics are e.g. radius, diameter, average path length, or assortativity coefficient. Those metrics can be extended to weighted networks and need to be interrelated to entropy production per node and network entropy production, respectively.

Centrality metrics. Centrality metrics quantify the 'importance' of nodes. Different variations exist like degree, (random walk) closeness, information, betweenness, or Eigenvector centrality like PageRank. Considering again the R mailing list, we observed a strong correlation between entropy production per node H_i and e.g. betweenness centrality. Nodes in the social network that have a high probability to occur on a randomly chosen shortest path between two randomly chosen nodes have a high betweenness. Hence, those nodes are also responsible for high entropy production in the network. In contrast, closeness centrality and entropy production revealed no correlation. Closeness measures how fast it will take to spread information from a single node to all other nodes sequentially.

Symmetry measures and entropy measures. Current developmentsintroduce measures of symmetry and their relation with measures like Graph entropy [12]. The concept of Graph entropy is based on a probability distribution on the node set V of the graph [13], $G = \sum_{i=1}^{|V|} p(v_i) \log \frac{1}{p(v_i)}$, but not on the ratio between incoming and outgoing message rates as for entropy production. Hence, graph entropy measures the amount of information within the graph based on $p(v_i)$. Symmetry of complex networks means invariance of adjacency of nodes under the permutations on the node set itself and a symmetry index is defined in [14]. This concept is close to entropy production, however, symmetry measure are defined on network topology only. In a similar way, the symmetry based structure entropy of complex networks [15] quantifies the heteogeneity of a network system based on automorphism partition of the node set into equivalent cells. Thereby, the probability that a node belongs to the cell is used while message rates are not part of this concept. A comparison of those measures with entropy production is relevant future work.

7 Conclusions

In the current Internet, social networks like Facebook gain more and more popularity and attract millions of users. In a social network the users are connected to each other and as a key feature social media platforms allow interactions between users like exchange of messages. In complex network research, the majority of existing quantities analyze the structural properties of the emerging network topology and the growth dynamics, respectively. For social networks, however, beyond the network topology the interaction among individuals needs to be characterized. Further, the dynamics of a stationary state of a social network is not uniquely given, rather there is a large variety of possible realizations. Hence, there is a gap in describing dynamical properties of social networks in stationary conditions, i.e., when the network topology as well as the probability for receiving and sending messages do not change in the long-term limit. Inspired from statistical physics, we introduce a quantity called entropy production to characterize the stationary properties of arbitrary social networks. The entropy production measures the directionality of the information exchange and vanishes for perfectly balanced communication. Defining the entropy production of a node as the sum over the entropy of all its links, one can identify nodes contributing preferentially to balanced or unidirectional information transfer. Hence, entropy production is a valuable measure for link and node analysis and rating and can be used to detect hidden structures and interactions in networks. Since the application of entropy production is not limited to social media network, but can be used for communication networks or interaction graphs in general, it can be applied for a variety of different purposes like anomaly detection [16] but also characterization of traffic flows in the Internet, e.g. for BitTorrent swarms [17]. Future work addresses the application of entropy production to such use cases but also to relate the quantity with centrality or symmetry measures for various network topologies and message exchange models.

References

1. Jin, E.M., Girvan, M., Newman, M.E.J.: Structure of growing social networks. Phys. Rev. E 64 (September 2001)
2. Barnes, N.G., Andonian, J.: The 2011 fortune 500 and social media adoption: Have america's largest companies reached a social media plateau? (2011), `http://www.umassd.edu/cmr/socialmedia/2011fortune500/`
3. Hoßfeld, T., Hirth, M., Tran-Gia, P.: Modeling of Crowdsourcing Platforms and Granularity of Work Organization in Future Internet. In: International Teletraffic Congress (ITC), San Francisco, USA (September 2011)
4. Schnakenberg, J.: Network theory of microscopic and macroscopic behavior of master equation systems. Rev. Mod. Phys. 48 (October 1976)
5. Schreiber, T.: Measuring information transfer. Phys. Rev. Lett. 85 (July 2000)
6. Andrieux, D., Gaspard, P.: Fluctuation theorem and onsager reciprocity relations. J. Chem. Phys. 121(13) (2004)
7. Seifert, U.: Entropy production along a stochastic trajectory and an integral fluctuation theorem. Phys. Rev. Lett. 95 (July 2005)
8. Zeerati, S., Jafarpour, F.H., Hinrichsen, H.: Entropy production of nonequilibrium steady states with irreversible transitions (2012) (under submission)
9. Box, G.E.P., Tiao, G.C.: Bayesian Inference in Statistical Analysis. John Wiley & Sons, New York (1973); (reprinted in paperback 1992 ISBN: 0-471-57428-7 pbk)
10. R Mailing Lists, `http://tolstoy.newcastle.edu.au/R/`
11. Ebel, H., Mielsch, L.I., Bornholdt, S.: Scale-free topology of e-mail networks. Phys. Rev. E 66 (September 2002)
12. Garrido, A.: Symmetry in complex networks. Symmetry 3(1) (2011)
13. Garrido, A.: Classifying entropy measures. Symmetry 3(3) (2011)
14. Mowshowitz, A., Dehmer, M.: A symmetry index for graphs. Symmetry: Culture and Science 21(4) (2010)
15. Xiao, Y.H., Wu, W.T., Wang, H., Xiong, M., Wang, W.: Symmetry-based structure entropy of complex networks. Physica A: Statistical Mechanics and its Applications 387(11) (2008)
16. Bilgin, C., Yener, B.: Dynamic network evolution: Models, clustering, anomaly detection. Technical report, Rensselaer University, NY (2010)
17. Hoßfeld, T., Lehrieder, F., Hock, D., Oechsner, S., Despotovic, Z., Kellerer, W., Michel, M.: Characterization of BitTorrent Swarms and their Distribution in the Internet. Computer Networks 55(5) (April 2011)

An Analysis of the Overlap of Categories in a Network of Blogs

Priya Saha and Ronaldo Menezes

Abstract. We live in a world where information flows very rapidly and people become aware of events on the other side of the world in a matter of seconds; this a consequence of the globalized, fully-connected world we live in. Information spreads via many different channels, but more recently we have witnessed the birth of the information-over-online-social-network phenomena. This means that more and more people get their news from online social networks such Facebook and microblogs such as Twitter. Yet, another source of information are weblogs (or blogs). Bloggers (people who write to blog or own a blog) are capable of influencing a lot of people and they even tend to be sources of information to mainstream news media. This paper delves into an issue relating to the ability of information to spread, but instead of tracking information itself, we look at the infrastructure that is in place linking blogs. We argue that the structure itself is an enabler or disabler of information spread depending on a categorization. This paper categorizes blogs and studies the level of overlap between these categories.

1 Introduction

How fast does information spread on today's connected world? Although this is hard to be measured, there have been attempts. Clark [7] has provided a table in his book that shows how fast information used to spread in the 19th century. In one of his examples he argues that it took 17 days for the information about the Battle of Trafalgar (1805) to arrive in London. He argues that since the battle took place approximately 1,100 miles away, the information travelled a mere 2.7 mph. The information about the assassination of Lincoln (1865) took 12 days to reach London from Washington DC leading to the conclusion that the information travelled around

Priya Saha · Ronaldo Menezes
BioComplex Laboratory, Department of Computer Sciences,
Florida Institute of Technology, Melbourne, Florida, USA
e-mail: psaha2010@my.fit.edu, rmenezes@cs.fit.edu

G. Ghoshal et al. (Eds.): *Complex Networks IV*, SCI 476, pp. 59–70.
DOI: 10.1007/978-3-642-36844-8_6

12 mph. If we apply Clark's approach to today's world the results are staggering. The 2008 Sichuan, China earthquake was made known on Twitter (in English) about 8 minutes after the event. This means that someone in London would have seen this soon after that. Since Sichuan is 5,100 miles away from London one can conclude that the information travelled at 38,000 mph!

Regardless of the numbers, what all the accounts above have in common is that information spread *through a medium*; there must be a framework in place that enables the spread. In the 19th century this structure was probably messagers carrying the information with them in the form of letters or even carrier pigeons sent back and forth. Today however the infrastructure consists of online social networks, blogs and micro blogs. There are many works on the spread of information in micro blogs [13, 12] and online social networks [4, 5] but not much discussed with regards to the hindrances that could be imposed by the structure itself.

1.1 Motivation

The study of information diffusion is an active research topic with many works available in the literature [13, 6, 12], yet these works do not address the issue of connectivity of the infrastructure where the information is stored. We are familiar with the work on the connectivity of the Web [2] and its characteristics, but in the context of blogs we have one extra layer, the "topic" of discussion of that blog. We pose a basic question here: how much hindrance can the topic be to the connectivity of the blogosphere? To answer this question we look at how these topics are connected and at what level.

Blogs form an integral part of World Wide Web. Though blogs are not as popular as other social networks such as Facebook and LinkedIn, they are an essential source of information to many people in the world. Bloggers collect news from all over the world and post them to their blogs to attract readers. The collection of blogs enables readers to read information from news-oriented blogs instead of going through news channels or newspapers.

The connectivity of blogs in the blogosphere (a network of blogs) can tell us a lot about the diffusion of information in networks because the structure is a necessary condition for the diffusion. The structure of the blogosphere refers to the links or the connections between blogs. Readers can reach a blog in two ways: either by using search engines, or by following links available in other blogs (hyperlinks).

We hypothesize that although blogs are organized in categories these categories connect well to each other to form a single network. Additionally we want to verify how organized the blogosphere is with regards to categories. We believe the organization should be inherent to the structure because it allow us to navigate between blogs more easily; it is logical to believe the blogs discussing politics, for instance, contain links to other politically-oriented blogs. If this is the case the blogosphere would have sections similar to a local newspaper. Newspapers contain information about many subjects including sports, politics and science. These subjects are very well organized and somewhat isolated from each other within the newspaper. Hence, it is very easy for a reader to find the information they are seeking. Additionally, the sections tend not to link to each other—one may read the sport section of a newspaper and never know what is happening in the political scene. Is this separation present in the blogosphere? and to what extent? This is the subject of this paper.

2 Related Works

There have been many works related to the blogosphere, although none looked at the issues of categories. Devezas et al. [9] studied blog features based on the popularity of links. They worked on the Portuguese blogosphere. The dataset was collected from SAPO, a Portuguese internet service provider. The study was aimed at understanding the characteristics of the Portuguese blogosphere such as the length of posts, and posting frequencies. There were about 70,000 blogs and more than 400,000 links in the network that was used. The study was also made to analyze the features of highly-cited blogs, less-cited blogs, and posts in blogs over the years. It was found that links of a blog with other blogs increase by 17.88% per month, demonstrating the role of blogs as information source to many people and also that these people carry information from blog to blog. It was found that the blogs that frequently form links with other blogs are highly cited. The evolution of new words or posts decrease when the blogs become less cited. This proved a relation between post length and citation. The blogs which have more content are more cited.

Cointet and Roth [8], in their work on social and semantic aspects of blogs, looked at the formation new connections among blogs. The statistics of a blog determines its cognitive attributes and the information that flow through it. By studying the cascade patterns of a blog, one can analyze its spread of influence. Cascade study considers chains of posts made by the bloggers in the path of information flow. The study was done on a dataset of the "PresidentialWatch08" project which is a small-to-medium sized network of 1,066 blogs and 229,736 total edges which consists of repeated and non-repeated edges. This study revealed that the textual contents that a blogger uses are mostly the issues he likes to discuss and hyperlinks are digital resources that are related to such discussion. It is further observed that the immediate neighborhood of a blog is semantically very close to the overall semantic distance between other pairs of blogs in the blogosphere. So a blogosphere has a strong homophily and dynamic nature. This work seems to indicate a tendency towards clustering around what we call "categories" but the connectivity among categories is not studied.

Qureshi et al. [15] wrote an algorithm to find clusters based on topics of discussions in the blogosphere. Topic cluster refers to blogs sharing the same interest. It analyzes hyperlinks and content in the blogosphere. Once the topics are identified, they are ranked according to their influence in the network. The metric which had been used to rank the topic based clusters is called "Topic Discussion Rank". Another metric, "Topic Discussion Isolation Rank," had been used to identify cluster. Usually bloggers of a particular interest discuss the topic of interest in their blog posts. They also get linked to other blogs of similar interests. Eventually they form a cluster of blogs of similar interests that discuss particular topics. Extracting bloggers who are interested in a particular area is of great importance in the information retrieval research. In this work, the interest of a blog was determined by three dimensions: occurrence of a particular word in the posts, number of posts, and parts of speech of the words. To analyze properly, a natural language processing technique had been used. For example, if the word "democracy" is searched, the adjective "democratic" and the adverb "democratically" were also considered.

3 Building Blog Networks

Recall the main issue we deal in this paper is the possible link between blog categories. In order to understand how this connectivity can take place we show a toy example in Figure 1. The figure shows the evolution of the connectivity between 3 blogs from 3 different categories: sports, politics and religion. In the figure, each column represent a snapshot in time. Initially we have 3 blogs with 3 bloggers writing posts to them. At this stage the three blogs do not link to each other. Now at step 2 (column 2 in Figur 1) we have "Blogger 2" read something on the sports blog and post about it on his politics blog along with a link to the original post in the sports blog; this is represented in the second column of Figure 1. After the post, the politics blog contains a hyperlink to the sports blog (column 3 in Figure 1). Also on column 3 we have "Reader 2" and "Reader 3" posting something to the sports blog that they read from politics and religion blogs respectively. After all is done the sports blog contains a link to the religion and the politics blog (fourth and last column in Figure 1. Although a simple example, we can see that over time, blogs may start connecting to each other. It is not a hard exercise to understand that what is described here can easily happen in the real blogosphere.

Next we had to decide on a dataset to work on. There are many blog hosting sites, such as Blogspot, Slashdot, WordPress, Gizmodo, and others. Among them, Blogspot and WordPress are the most common. We decided to go with Blogspot because we believe it to be the most diverse and Google has a good policy of removing spam blogs from their servers. We built a network from Blogspot, where the nodes are blogs and the edges are the connections between the blogs. Most blogs have connections to favorites, and many of them are blogs hosted on the same domain. Our network is built from these links. Figure 2 shows an example of these links. "Hyperbole and a Half" is one of the best known blogs on Blogspot. In Figure 2 we show a section of the blog with links to other places (also blogs).

We assumed that if two blogs connect, there is a chance of information exchange. The first step was to collect our dataset. We started by defining some categories. We chose 12 categories that reflect general categories of newspapers or TV programs: arts, business, cooking, education, finance, health, music, politics, religion, science, sports, and travel. We then used Google itself to give us a list of the top blogs in each category. We collected the top 10 blogs of each category from Google using the search string `site:Blogspot.com category` where `category` is one of the 12 listed above. We explored the blogs at different *depths*. By depth, we mean

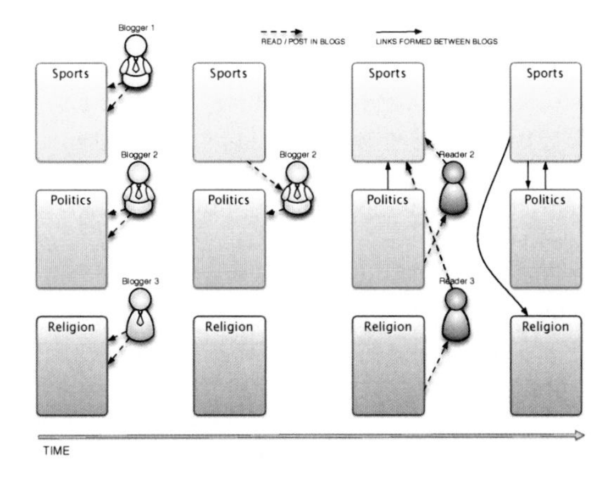

Fig. 1 The above network is an example of the formation of the blog network and the evolution of links between blogs as a function of time. Bloggers are the people who write blogs and readers are the people who read blogs (may also be bloggers). Readers my comment on posts placed by bloggers and with this action add hyperlinks in one blog to another (last column in the figure).

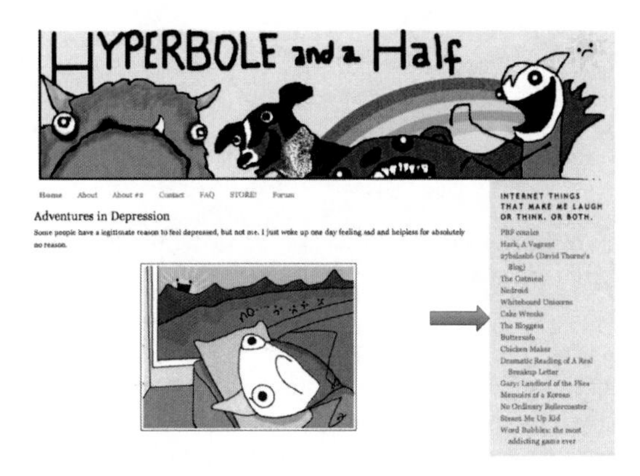

Fig. 2 Blog "Hyperbole and a Half" and its section with links to other locations including other blogs. The links are shown by the red arrow.

the distance from the top 10 predefined blogs. We considered the top 10 blogs of each category as depth 0. The blogs to which depth-0 blogs link are considered as depth-1 blogs. Similarly, the depth-1 blogs link to blogs at depth 2. Our study explored the blogosphere (network of blogs) until depth 4. One can make a general assumption that since the categories are independent of each other as we see in newspapers or TV channels, there may be little overlap between the categories.

We created a Web crawler that, starting from the top 10 blogs given by the Google search above, traverses the links all the way to depth 4 and link blogs as it crawls. If the crawler finds a blog name that ends in blogspot.com in the body text of the current blog, a link will be created from the blog being crawled to the other blog found in the text. The new blogs that are collected, are in turn crawled and linked to the blogs that are found in their posts, until we reach the desired depth. During the crawling process, we also assigned the initial category as an attribute of every reached site. That is, if a cooking blog is the starting point of the blog, "cooking" will be used as an attribute of that blog and every blog reached from it. We use this approach to measure the level of reach and overlap between the initial categories.

This assignment of categories to reach blogs is quite important to be understood. Figure 3 tries to better explain what is taking place. The figure shows a sample with 3 starting points shown in colors. These starting points are categorized as Business (B), Cooking (C), and Music (M). Once they are crawled the reached blogs are categorized. For instance a blog with (M,C) indicates that it has been reached by a music blog (M) and by a cooking blog (C) and hence is an overlap between these two categories.

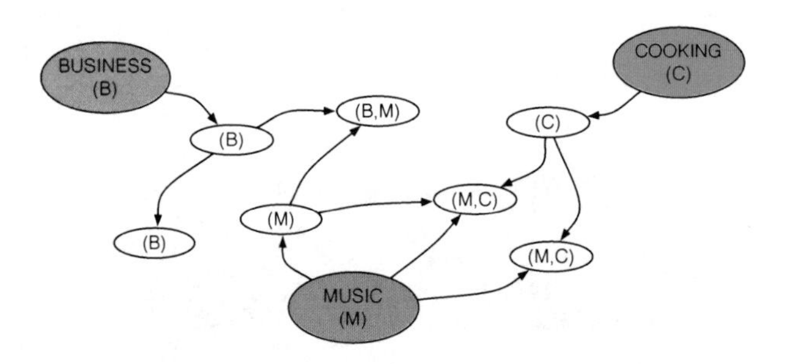

Fig. 3 Picture depicts a simple example of what is done during the crawling process. Initial blogs are pre-categorized and that category is transfered to every other blog they reach.

Our networks are directed (as shown in Figure 3) because direction represents the true nature of the hyperlinks. Our networks are also weighted because it can help us discover the highly interconnected nodes in the network. If an edge between two blogs has a high weight, it means that there may be heavy exchange of information between the blogs. The weight comes from the number of links of one blog found in the currently-crawled blog. We crawled the blogs of the 12 categories and built a network for each category. We studied the structures of all the networks to find out if their characteristics differ. Regardless of the categories or depths, all our networks show similar characteristics. We will discuss the analysis and results of our experiments in Section 4.

While crawling the blogosphere, we came across many blogs that are common to more than one category; the situation shown in Figure 3 was common. We refer the common blogs as *overlap* blogs in our study. Overlap blogs are of particular interest to us because they are the ones who play active roles in diffusion of information over the barriers of categories. As we crawled Blogspot at increased depths, we noticed that the overlapped blogs start connecting with each other forming a network within a network. We concentrated on the overlapped blogs and extracted them from the full network. The resulting network of only overlapped blogs is also very well connected. We named this network of overlap blogs as the *sink network.*

4 Experimental Results

We examined primarily the overlap of blog categories but in this process we also looked at the reachability of each category, and looked at the network properties for each category, full, and sink networks.

4.1 Category Reachability

Table 1 shows the number of blogs we can reach at every depth. The reachability of music blogs is really high at any depth; from 10 initial nodes, music blogs can reach more than 190 thousand blogs at depth 4. The table also shows that the number of blogs in the sink network has a significant increase from depth 3 to depth 4, almost like in a phase transition. We believe this is because the sink structure starts connecting with other nodes in the network to form the giant component in the network. Although further work would be necessary, we think that the high change is a sign that in a few more depths one should be very close to having the giant component equal to the size of the entire network. In our analysis, the giant component starts to emerge in depth 3 and grows significantly at depth 4. We have not crawled depth 5 because of lack of resources, at that level we need to wait for days to extract the links since the number of blogs are in the millions.

We assume that we can reach every blog of the blogosphere at a certain depth. At depth 3, the sink consists of about 13% of the full network and at depth 4, the sink network consists of about 35% of the full network. Hence, the sink network seems to converge towards the full network.

4.2 Network Measures

We have looked at some network measures for the networks we dealt with in this paper; the results are presented in Table 2. Many of these measures are well-known in the literature. Newman [14] provides an extensive description of these measures and how they are computed.

The metrics in Table 2 were calculated for the networks at depth 3 because the tools used do not perform well with larger networks. However, the purpose of Table 2 is solely to characterize the networks; at larger depths these values are likely to be similar given the growth of the network follow the same process throughout the Blogosphere.

The degree distributions of all the blog networks (categories, full, and sink) follow a power-law distribution indicating that majority of the nodes in the networks have low degree whereas few nodes have high degree. An interesting feature of all

Table 1 Number of blogs that are reachable at every depth. We can see that the number of blogs increase with the increase in the depth crawled. The full network is the total number of blogs reachable in the Blogspot blogosphere. The number of blogs in the categories includes are only the ones that can be reached by the initial nodes in that category alone.

Categories	Depth 0	Depth 1	Depth 2	Depth 3	Depth 4
Arts	10	168	1,240	11,359	90,962
Business	10	45	173	240	10,182
Cooking	10	84	1,005	9,711	77,760
Education	10	76	514	4,301	34,076
Finance	10	22	94	361	1,252
Health	10	73	634	7,537	96,447
Music	10	**527**	**6,783**	**38,120**	**194,607**
Politics	10	52	639	5,625	63,326
Religion	10	32	410	4,344	22,623
Science	10	187	1,376	11,416	90,842
Sports	10	73	601	4,472	40,598
Travel	10	51	344	2,568	28,557
Sink network	0	17	506	11,233	148,218
Full Network	120	1,373	13,307	84,701	426,923

Table 2 We have considered 12 different categories and formed network of each category before merging them to form the full network. Here, n is the total number of nodes in the network, m is the total number of edges in the network, λ_{out} and λ_{in} are the exponent of the power-law distribution for the out- and in-degree [2], ℓ is the average path length, γ is the scaling law coefficient [10] and C is the average clustering coefficient.

Network	n	m	λ_{in}	λ_{out}	ℓ	γ	C
Arts	11,359	16,650	2.47	1.10	4.05	1.43	0.05
Business	240	345	2.11	0.96	2.30	1.28	0.06
Cooking	9,711	17,346	2.38	1.08	3.98	1.22	0.06
Education	4,301	6,508	2.07	0.86	3.63	1.20	0.06
Finance	361	522	2.33	0.73	2.55	1.30	0.07
Health	7,537	10,115	2.26	0.60	3.87	1.72	0.04
Music	38,120	118,720	2.05	1.45	4.58	0.76	0.08
Politics	5,625	9,026	2.13	0.78	3.87	1.25	0.05
Religion	4,344	6,633	1.78	1.02	3.63	1.20	0.04
Science	11,416	19,507	2.18	1.11	4.05	1.41	0.06
Sports	4,472	6,775	2.31	1.06	3.65	1.43	0.08
Travel	2,568	3,091	2.64	0.92	3.40	1.28	0.03
Sink Network	11,233	35,898	2.18	1.40	4.05	0.95	0.16
Full Network	84,701	206,373	2.30	1.59	4.92	0.99	0.07

the blog networks is that out-degree distributions follow the power law with exponential cut-off (GPL-EC) [1]. When created, a blog tries to establish connections with other existing blogs to become popular. With time, the blog stops connecting to others causing theout-degree distributions to be capped, thus the cut-off. Figure 4 shows an example of the distribution for the case of the sink network at depth 3.

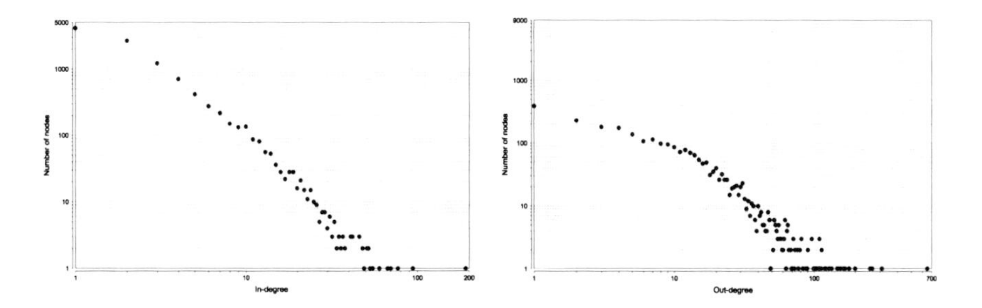

Fig. 4 In-degree (left) and out-degree (right) distribution for the nodes in the sink network at depth 3. The out-degree distribution is follows the power law with exponential cut-off.

If we look at all the blog networks (category networks, full and sink networks) we can argue that they are all small world, hierarchical and scale free. The small-world characteristic (given by small ℓ and high C values means that information flowing following the links of the blogs can reach other blogs in a few steps. The hierarchical property (given by the fact that $\gamma \sim 1$ in most networks) tells us that the blogosphere probably was built bottom up with similar blogs connecting forming clusters which in turn connect themselves in a sort of hierarchical manner. Last the scale-free property indicates the existence of hubs in the network.

4.3 Category Overlap

Recall that our main purpose was to look at the formation of the sink network and the level of overlap between categories. As we argued before, the sink network starts to become connected (overlap nodes start to point to other overlap nodes) soon after we reach depth 1. At depth 2, the sink network already shows high connectivity (see Figure 5).

However we can see that Figure 5 does not tell us what category is overlapping with what. For this we need to do a pairwise comparison of the categories.

In Table 3, we show the number of blogs in common in every two categories. The way to use the table is first to understand that because of the symmetry we are showing only half of the table. Then to understand the overlap of one particular category, we need to start at the column of that category until we reach the arrow and then follow the arrow on that line going to the right-hand side. As one example of the above we highlight on the table the information related to Education in **boldface**.

Table 3 allows us to see the proximity between categories. For instance, if we concentrate on Table 3 we can clearly see that Business blogs are more similar to Science than they are to Education; similarly, Arts blogs are more similar to Music blogs than Sports. One interesting overlap is the overlap of Arts and Science. Although it is hard to be sure of the reasons, we know that many scientists are aware of the power of visualizations to present their results [11, 3].

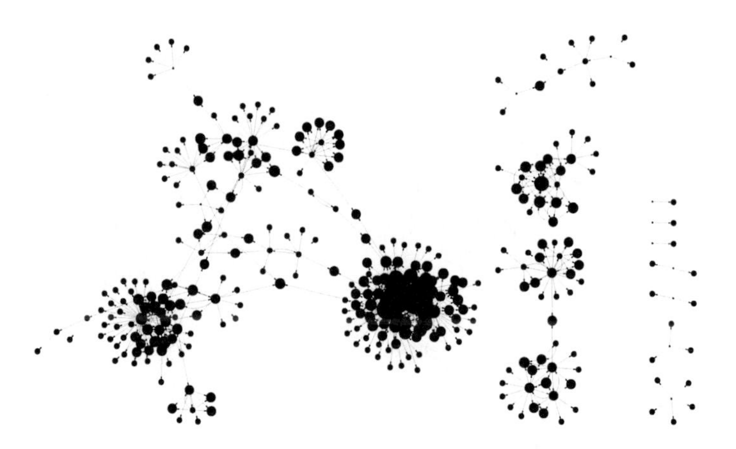

Fig. 5 Snapshot of the overlap nodes forming the sink network at Depth 2 of the crawling

Table 3 The table shows the number of blogs in common to every pair of categories at depth 4. The results show a huge overlap between Music and Arts categories which is somewhat expected. However one can see also interesting issues such as the overlap of Science and Arts. The information regarding Education is highlighted in **boldface**. Note that we have to use the column and the row of Education exemplify the way to get the pairwise comparison between Education and all other categories.

Categories	Arts	Business	Cooking	**Education**	Finance	Health	Music	Politics	Religion	Science	Sports	Travel
Arts	↪	5,591	14,804	**17,562**	597	27,755	40,257	22,621	7,140	30,756	14,395	10,951
Business		↪	1,598	**2,715**	224	3,961	4,078	3,483	1,163	4,546	1,839	2,417
Cooking			↪	**6,304**	153	19,547	15,660	12,495	7,151	21,132	7,227	11,870
Education				↪	**406**	**14,680**	**15,946**	**15,638**	**4,186**	**16,434**	**10,734**	**4,349**
Finance					↪	484	418	542	140	431	408	206
Health						↪	26,128	27,974	8,074	35,313	15,324	11,359
Music							↪	20,695	5,997	25,592	15,814	9,649
Politics								↪	7,310	27,815	16,182	8,730
Religion									↪	11,935	5,311	3,744
Science										↪	15,472	13,930
Sports											↪	4,521
Travel												↪

The overlap in Table 3 demonstrates that even for blogs with low reach, the infrastructure is in place to help the difusion of information. This means that we do not seem to see segregation of categories in the blogosphere in the same way we see in traditional medial. Recall that we argued that newspapers are organized in such a way that information about a particular category rarely reference information from other parts of the newspaper. This does not seem to be the case in the Blogosphere where we see the emergence of a large sink network (in relation to the total number of nodes) and a significant overlap between any pair of categories.

5 Conclusion

In this paper, we performed experiments on the connectivity of Blogspot, Google's blog hosting service. We used the blogspot because it is one of the most commonly-used hosting services. The experiments that we did on the blogspot include: *(i)* reachability of the blogs; *(ii)* measures of the networks; and *(iii)* the structure category overlaps in the blogosphere. We built networks from the blogspot and examined their characteristics. While doing this analysis, we found many blogs can be reached from more than one category. These blogs play active roles in information diffusion because they are related to more than one category. Blogs that can be reached from more than one category form overlaps between categories. We referred to this overlapped structure as the sink network.

We limited our work to 12 categories that are of general interests to people. At the first step, we used Google to get the names of top 10 blogs. We looked into the reachability in the blogosphere per category starting from the 10 initial blogs. To analyze the reachability, we crawled the blogosphere to pre-defined depths. During this process, we considered the category of a blog to be the same as the blog from which the crawler has started. Our results show that music blogs have the highest reach. This agrees with the fact that many of the popular social media site have music as their number one category of interest, for instance a quick search on YouTube for the most watched videos give us that 8 out of the top 10 most watched videos are related to music.

There are millions of blogs on Blogspot. In our study, we found that the sink forms a major part of the blogosphere as the depth increases. We would like to see at which depth the sink becomes the entire blogosphere, or in other worlds, when the sink network will percolate the entire blog network. This is not an easy process because there are mechanisms that Google puts in place to avoid us running crawlers. In practice, it means that the data collection for further depths would take a fairly long time. Yet, a couple more depths could tell us a little more about the trends in the sink network.

Our study reveals that categories cannot limit the flow of information in blogs. We would like to consider other factors (like topography or language) and examine if they play any role in the information diffusion. As the world becomes more globalized it may be possible that language is also not a barrier of the connectivity.

References

1. Amaral, L.A., Scala, A., Barthelemy, M., Stanley, H.E.: Classes of small-world networks. Proc. Natl. Acad. Sci. 97(21), 11149–11152 (2000)
2. Barabási, A.L., Albert, R.: Emergence of scaling in random networks. Science 286(5439), 509 (1999)
3. Bastian, M., Heymann, S., Jacomy, M.: Gephi: An open source software for exploring and manipulating networks. In: International AAAI Conference on Weblogs and Social Media (2009)

4. Brown, J., Broderick, A.J., Lee, N.: Word of mouth communication within online communities: Conceptualizing the online social network. Journal of Interactive Marketing 21(3), 2–20 (2007)
5. Centola, D.: The spread of behavior in an online social network experiment. Science 329(5996), 1194–1197 (2010)
6. Cha, M., Mislove, A., Gummadi, K.P.: A measurement-driven analysis of information propagation in the flickr social network. In: Proceedings of the 18th International Conference on World Wide Web, WWW 2009, pp. 721–730. ACM (2009)
7. Clark, G.: A Farewell to Alms: A Brief Economic History of the World. Princeton University Press (2008)
8. Cointet, J.-P., Roth, C.: Socio-semantic dynamics in a blog network. In: Proceedings of the 2009 International Conference on Computational Science and Engineering, CSE 2009, vol. 04, pp. 114–121. IEEE Computer Society, Washington, DC (2009)
9. Devezas, J.L., Ribeiro, C., Nunes, S.: Studying blog features over link popularity. In: Proceedings of the First Workshop on Social Media Analytics, SOMA 2010, pp. 31–34. ACM, New York (2010)
10. Dorogovtsev, S.N., Goltsev, A.V., Mendes, J.F.F.: Pseudofractal scale-free web. Phys. Rev. E 65(6), 066122 (2002)
11. Heer, J., Boyd, D.: Vizster: Visualizing online social networks. In: Proceedings of the 2005 IEEE Symposium on Information Visualization, INFOVIS 2005, pp. 32–39. IEEE Computer Society (2005)
12. Kwak, H., Lee, C., Park, H., Moon, S.: What is twitter, a social network or a news media? In: Proceedings of the 19th International Conference on World Wide Web, WWW 2010, pp. 591–600. ACM (2010)
13. Leskovec, J., Backstrom, L., Kleinberg, J.: Meme-tracking and the dynamics of the news cycle. In: Proceedings of the 15th ACM SIGKDD International Conference on Knowledge Discovery and Data Mining, KDD 2009, pp. 497–506. ACM (2009)
14. Newman, M.: The structure and function of complex networks. SIAM Review 45(2), 167–256 (2003)
15. Qureshi, M.A., Younus, A., Saeed, M., Touheed, N., Pianta, E., Tymoshenko, K.: Identifying and ranking topic clusters in the blogosphere. In: Proceedings of the 2nd Workshop on The People's Web Meets NLP: Collaboratively Constructed Semantic Resources, Beijing, China, pp. 55–62 (August 2010)

The Relative Agreement Model of Opinion Dynamics in Populations with Complex Social Network Structure

Michael Meadows and Dave Cliff

Abstract. Research in the field of *Opinion Dynamics* studies how the distribution of opinions in a population of agents alters over time, as a consequence of interactions among the agents and/or in response to external influencing factors. One of most widely-cited items of literature in this field is a paper by Deffuant *et al.* (2002) introducing the *Relative Agreement* (RA) model of opinion dynamics, published in the *Journal of Artificial Societies and Social Simulation* (JASSS). In a recent paper published in the same journal, Meadows & Cliff (2012) questioned some of the results published in Deffuant *et al.* (2002) and released public-domain source-code for implementations of the RA model that follow the description given by Deffuant *et al.*, but which generate results with some significant qualitative differences. The results published by Meadows & Cliff (2012) follow the methodology of Deffuant *et al.* (2002), simulating a population of agents in which, in principle, any agent can influence the opinion of any other agent. That is, the "social network" of the simulated agents is a *fully connected* graph. In contrast, in this paper we report on the results from a series of empirical experiments where the social network of the agent population is non-trivially structured, using the stochastic network construction algorithm introduced by Klemm & Eguiluz (2002), which allows variation of a single real-valued control parameter μ_{KE} over the range [0,1], and which produces network topologies that have "small world" characteristics when μ_{KE}=0 and that have "scale free" characteristics when μ_{KE}=1. Using the public-domain source-code published in JASSS, we have studied the response of the dynamics of the RA model of opinion dynamics in populations of agents with complex social networks. Our primary findings are that variations in the clustering coefficient of the social networks (induced by altering μ_{KE}) have a significant effect on the opinion dynamics of the

Michael Meadows · Dave Cliff
Department of Computer Science, University of Bristol, Bristol BS8 1UB, U.K.
e-mail: michaeljmeadows86@gmail.com, dc@cs.bris.ac.uk

G. Ghoshal et al. (Eds.): *Complex Networks IV*, SCI 476, pp. 71–79.
DOI: 10.1007/978-3-642-36844-8_7

RA model operating on those networks, whereas the effect of variations in average shortest path length is much weaker.

1 Introduction: Opinion Dynamics and the RA Model

Although "opinion dynamics" has come to cover many fields ranging from sociological to physical and mathematical phenomena (Lorenz 2007), this paper will focus its examination on the Relative Agreement (RA) model, first presented by Deffuant *et al.* (2002).

Consider a group of n experts each with an opinion represented by a real a number x, marking a point on some continuum and an uncertainty u about their opinion. Random paired interactions will then take place with the experts potentially updating their own opinion and uncertainty, provided that there is a degree of overlap between the agents' opinion range given by their opinion +/- their uncertainty. These interactions will iterate until the entire population reaches a stable state. As would be intuitively expected, the population converges into stable clusters within the range of the initially maximal and minimal opinions in the population, with the number of the clusters being linked to the uncertainty applied to the agents. This is an outline of the RA model in its most basic form.

Suppose now that we add "extremist" agents to the population, defined as individuals having extreme value opinions and very low uncertainties. With the inclusion of extremists it can be found that the population could tend towards three main outcomes: *central convergence*, *bipolar convergence* and *single extreme convergence*. In central convergence, typical when uncertainties are low, the majority of the population is clustered around the central, "moderate" opinion. When uncertainties are raised, the moderate population could split into two approximately equal groups one of which would tend towards the positive extreme and the other towards the negative: referred to as *bipolar convergence*. With initial uncertainties set very high an initially moderate population can, on occasion, tend towards a single extreme (and hence is known as *single extreme convergence*).

To classify population convergences, Deffuant *et al.* (2002) introduced the y metric, defined as: $y = p'^2_+ + p'^2_-$ where p'_+ and p'_- are the proportion of initially moderate agents that have finished with an opinion that is classified as extreme at the positive and negative end of the scale respectively. Thus, central, bipolar and single extreme convergences have y values of 0.0, 0.5 and 1.0, respectively.

Meadows & Cliff (2012) recently demonstrated that while all three population convergences can indeed occur using various parameter settings, the specific parameter values that do allow for the more extreme convergences are not as originally reported by Deffuant *et al.* (2002). Meadows & Cliff state that they reviewed over 150 papers that cite Deffuant et al. (2002) but not a single one of them involved an independent replication of the original published RA results. In attempting to then replicate the results reported by Deffuant *et al.* (2002), Meadows & Cliff (2012) found a significantly different, but simpler and more intuitively appealing, set of results: as agents' initial uncertainty u increases, the instability of the population rises with it, resulting in a higher average y value; also, as the proportion of initially extremist agents increases, there is again a corresponding rise in the resultant instability. Furthermore, when there is a higher level of instability in the population, there is a greater chance of the population converging in a manner other than simply to the centre. Also, there is no region of (p_e, u) parameter space (where p_e is the proportion of the extremist agents in the population) area in which single extreme convergence is *guaranteed*, although there is a large area of parameter space in which that can occur. This makes intuitive sense because a population with an initial instability that would allow single extreme convergence must surely also be unstable enough to allow central and bipolar convergences.

2 Populations with Complex Structure in Their Social Networks

In their initial publication on the RA model, Deffuant *et al.* (2002) explored the opinion dynamics exhibited by populations on a fully connected graph, but it is clear that real social interactions generally do not operate in this manner. In a subsequent paper, Amblard & Deffuant (2004) explored the dynamics of populations in which the social networks had small-world topologies. However, Meadows & Cliff's (2012) recent publication of data that calls into question the results presented by Deffuant *et al.* 2002, also indirectly casts serious doubt on the results presented by Amblard & Deffuant (2004) – yet the Meadows & Cliff (2012) paper does not cite or explicitly discuss Amblard & Deffuant's 2004 paper, which in hindsight is something of an oversight. Therefore, in this paper, we use the public-domain Java code published as an appendix to Meadows & Cliff (2012) to explore issues similar to those studied by Amblard & Deffuant (2004), examining the dynamics of opinion influence and population-level convergence in groups of agents where there is complex nontrivial structure in the social network. We explore dynamic behaviours over a range of different social network structures, and for that reason, it is prudent to first present a very brief summary of them.

Small World (SW) networks were introduced by Watts & Strogatz (1998), and they are sufficiently well known that we presume they need no detailed introduction in this paper. SW networks exhibit *both* low average path lengths *and* social clustering. Watts & Strogatz introduced an attractively simple stochastic algorithm for constructing SW networks. Nevertheless, one limitation of SW networks as models of human social networks is the extent to which SW networks have unrealistic degree distributions. In real social networks, the majority of nodes often have few connections, while a small number have very high degrees. A well-known potential resolution of this was proposed by Barabási & Albert (1999). The Barabási-Albert (BA) algorithm could construct random graphs with low average path lengths that also obeyed a power law in degree distributions (*scale free networks)*. However, the BA model is unable to generate networks that exhibit clustering levels as high as those in observed social networks and so, although both SW and BA models were useful as research tools, neither could claim to be entirely realistic.

To construct a graph that would exhibit all three qualities observed in real social networks (short average path lengths, high clustering, and a power-law degree-distribution) algorithms have been developed that produce hybrid networks that mix SW and BA characteristics. The KE algorithm introduced by Klemm & Eguíluz (2002) is the one used in this paper. The KE model begins by taking a fully connected graph of size m, the nodes of which are all initially considered *active*. A network is then "grown" by adding nodes iteratively to all of the currently active nodes in the graph after which a random active node is deactivated and the newest node is assigned to be active. When adding these nodes however, with a probability μ each new connection the node forms is assigned to a node using preferential treatment (a node with a higher degree is more likely to be randomly chosen) as in the BA model. With this addition, we see that when μ=1.0 the resulting network is identical to the BA model and with μ=0.0 the network is generated with topological characteristics as in the SW model. As Klemm & Eguiluz (2002) note, for values of μ between 0.0 and 1.0, KE networks exhibit properties that are "hybrid" mixes of the properties of SW and BA networks. For that reason, in this paper we use KE networks to explore the dynamics of the RA model in nontrivially structured populations. Since the RA model and KE network both employ a parameter called μ, hereafter we'll refer to these as μ_{RA} and μ_{KE} respectively, to avoid confusion.

In this paper, we present very brief results from a systematic empirical exploration of the dynamics of RA interactions within a range ($0<=\mu_{KE}<=1$) of complex-structured populations, to explore the factors that influence stability of opinion distributions in the population as a whole. Space constraints present us from presenting here a full set of our results. Instead, in the next section we show a small selection of illustrative examples: for a more comprehensive account, see (Meadows & Cliff, forthcoming 2013).

3 Complex-Structured RA

3.1 Additional Parameters

In exploring the dynamics of the RA model with extremists in structured populations, we found that the population stability (i.e., how reliably a certain type of convergence was exhibited) was affected by a number of additional factors that have to be considered because of the added complexity of the social networks involved, which we discuss in more detail in the following paragraphs.

For example, we conducted an A-B comparison of how our experiment results were affected when the extremists were assigned to the population in two ways: *first* and *last*. The nature of the KE-net construction algorithm is such that when the agents in the population are listed in the order in which they were added to the graph, those first in the list have, on average, a significantly higher degree than those last in the list. Thus, if we assign extremists first, they will have the highest degree and if we assign them last they will have the lowest. By comparing these two assignment protocols we can see if the degree has a notable effect on the final population opinion. It is clear however that the degree alone is not be the sole significant factor: we should also consider the typical number of connections that an extremist agent has to a moderate agent (i.e. the *average extremist degree*, excluding non-influencing connections between extremists). For example, when we use parameter-values comparable to those used in previously published work (e.g. n=200, p_e=0.1 using clustering values of μ_{KE}=0.5 and m=3) we find that the average number of connections an extremist has to moderates is 5.1 when assigned first, but only 1.8 when assigned last.

3.2 Explorations of the RA Model on Complex Social Networks

3.2.1 Comparing Seeding Methods: Varying Extremist Connectivity

An explanatory illustration of the variation in end-of-run average y values that we found as we varied μ_{KE}, with the extremists assigned first and last, is illustrated in Figure 1, which uses coloured "heat maps" to shows the mean and standard deviation in y-values, for extremists assigned first (left hand column) and last (right-hand column). Following that, Figure 2 shows a more comprehensive array of such heat-map plots, illustrating the effect of varying μ_{KE} over its entire range.

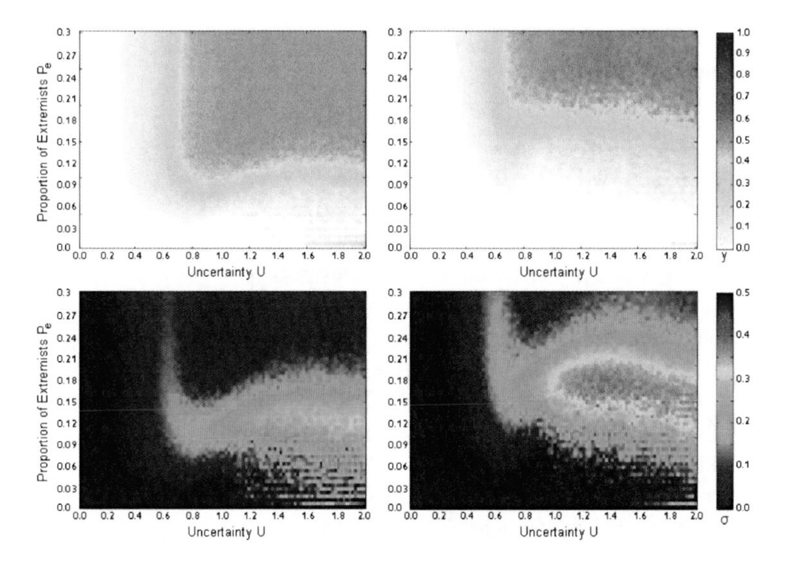

Fig. 1 Average y values (upper plot) and standard deviation (lower plot) for m=6, n=200, and μ_{KE}=0.6. Extremists assigned first (left) and last (right).

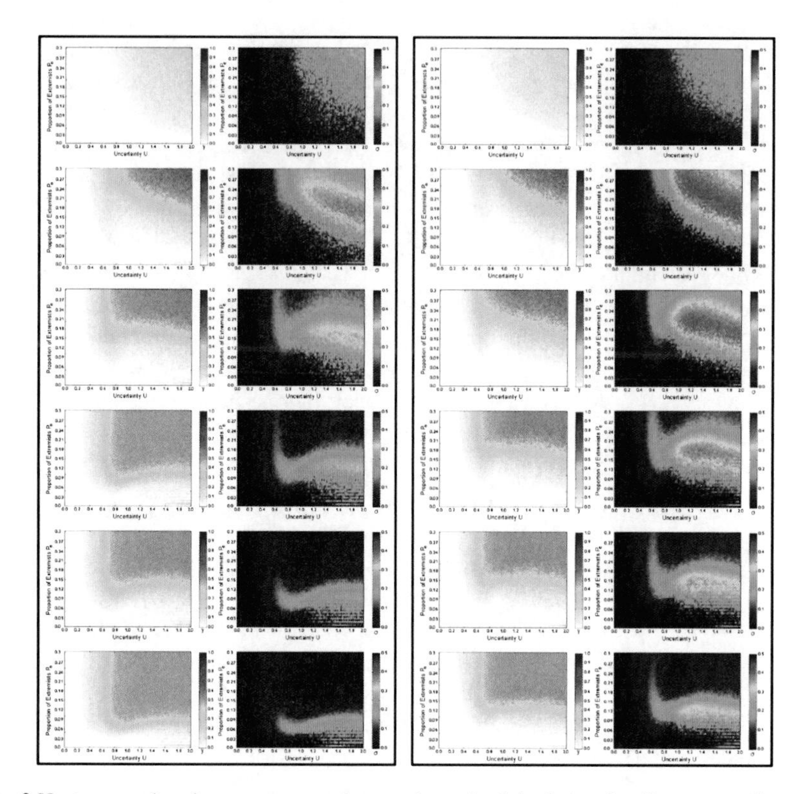

Fig. 2 Heat-maps showing average y values and standard deviation for (from top to bottom) μ_{KE} = 0.0, 0.2, 0.4, 0.6, 0.8, 1.0, when extremist agents are assigned first (left hand box) and last (right hand box), m=6, n=200. Heat-map colour-codes, and horizontal and vertical axis ranges, are the same as for Figure 1.

First, it can be seen in Figure 1 that populations where the extremist agents are assigned first have a higher average population instability than the equivalent population with extremists assigned last. This effect manifests itself on the y value heat-maps with larger areas of increased y values. With extremists assigned first, we see a pattern very similar to that in Meadows & Cliff (2012).

Somewhat counter intuitively however, we see when extremists are seeded last (leaving them with a lower average degree) the population is significantly less stable and shows a larger area of greater variance. This is the case because there are times when the extremists influence a greater number of the moderate agents, as there will be fewer connections between the extremist agents.

We find that all significant conclusions drawn from results where extremists are assigned first apply equally to populations where the degree of extremist agents is the population average (i.e., random assignment) or lowest (i.e., assigned last).

We have also explored the effects of varying network-size parameter n, and the effects of varying the SW/SF "mixing parameter" μ_{KE}. Full details of these studies are presented in (Meadows & Cliff, forthcoming 2013), but in brief our findings are as follows.

3.2.2 Effects of Network-Size *n*

As was shown in Meadows & Cliff (2012) increasing n causes an increase in stability in the initial population. This effect is intensified because the typical agent degree decreases proportional to n, thus the influence of the extremists becomes diluted as it propagates through the network and so the extremist agents struggle to firmly influence the opinions of the initially moderate agents.

3.2.3 Effect of the SW/SF "Mixing Parameter", μ_{KE}

We see the broadest range of results when μ_{KE} is around 0.3, lower than the focus of discussion of seeding. It is curious that the range that allows for such population instability appears to be quite small and further examination and explanation is presented in Meadows & Cliff (forthcoming 2013).

It is also of note to observe that as μ_{KE} increases, u may be set to smaller values and still allow for population instability. When μ_{KE} is 0.0 it is impossible to see anything other than central convergence but once the clustering coefficient lowers significantly we see results very similar to those in Meadows & Cliff (2012).

3.2.4 Overall Summary

To conclude this study of the effect of variation in the agents' social network's topological characteristics, via alterations in the value of μ_{KE}, Figures 3a and 3b summarise (with massive loss of subtlety) the data we have generated for the heat-map plots in Figure 2, and for similar plots not shown in this paper because

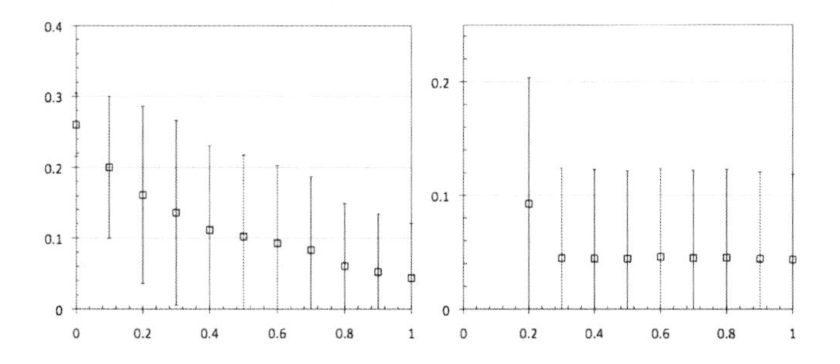

Fig. 3a and 3b 3a (left): Response in average y value (vertical axis: markers indicate mean; error bars at plus and minus one s.d.) resulting from variations in μ, expressed here on the horizontal axis as relative clustering coefficient $C(\mu)/C(0)$ in the normalised range [0,1]. 3b (right): response in average y value (vertical axis: markers indicate mean; error bars at plus and minus one s.d.) resulting from variations in μ, expressed here on the horizontal axis as relative average shortest path length $L(\mu)/L(0)$ in the normalised range [0,1].

of length constraints, for the exploration of systematic variations in n and μ_{KE}. In Figures 3a and 3b we show the average y-values and associated σ_y as functions of the clustering coefficient and shortest path length respectively, and it can quite clearly be seen that the mean y-value and is most heavily dependent on the clustering coefficient, when compared against shortest path length. We see in Figure 3a that as the shortest path length increases, we do not see any notable change in the average values of the population, yet in Figure 3a we see a marked decreased in average y values for the population as the clustering coefficient increases.

4 Conclusion

The RA model operating in a clustered population is a better model of real social systems and offers a significant potential for research into opinion dynamics. It is also clear that although the properties of this new system are beginning to be uncovered, there is already a great deal of scope for adding extra realism to the model.

References

Amblard, F., Deffuant, G.: The role of network topology on extremism propagation with the Relative Agreement opinion dynamics. Physica A 343, 725–738 (2004)

BarabáSi, A., Albert, R.: Emergence of Scaling in Random Network. Science 286, 509–512 (1999)

Brehm, S., Brehm, J.: Psychological Reactance: a theory of freedom and control. Academic Press, New York (1981)

De Groot, M.: Reaching a Consensus. J. Am. Statis. Assoc. 69(345), 118–121 (1974)

Deffuant, G., Neau, D., Amblard, F.: Mixing beliefs among interacting agents. Advances in Complex Systems 3, 87–98 (2000)

Deffuant, G., Amblard, F., Weisbuch, G., Faure, T.: How can extremism prevail? J. Artificial Societies and Social Simulation 5(4), 1 (2002)

Deffuant, G.: Comparing Extremism Propagation Patterns in Continuous Opinion Models. Journal of Artificial Societies and Social Simulation 9(3), 8 (2006)

Friedkin, N.: Choice Shift & Group Polarization. Am. Socio. Rev. 64(6), 856–875 (1999)

Hegselmann, R., Krause, U.: Opinion dynamics and bounded confidence: models, analysis and simulation. J. Artificial Societies and Social Simulation 5(3), 2 (2002)

Klemm, K., Eguíluz, V.: Growing Scale-Free Networks with Small-World Behavior. Physical Review E 65, 057102 (2002)

Krause, U.: A Discrete Nonlinear and Non-Autonomous Model of Consensus Formation. In: Proc. 4th Int. Conf. on Difference Equations, August 27-31, pp. 227–236 (1998-2000)

Lorenz, J., Deffuant, G.: The role of network topology on extremism propagation with the relative agreement opinion dynamics. Physica A 343, 725–738 (2005)

Lorenz, J.: Continuous Opinion Dynamics under Bounded Confidence: A Survey. International Journal of Modern Physics C 18, 1–20 (2007)

Meadows, M., Cliff, D.: Reexamining the Relative Agreement Model of Opinion Dynamics. Journal of Artificial Societies and Social Simulation 15(4), 4 (2012)

Meadows, M., Cliff, D.: The Relative Agreement model of opinion dynamics in populations with complex social network structure (forthcoming, 2013) (manuscript)

Watts, D., Strogatz, S.: Collective Dynamics of Small-World Networks. Nature 393, 440–442 (1998)

Capturing Unobserved Correlated Effects in Diffusion in Large Virtual Networks

Elenna R. Dugundji, Ate Poorthuis, and Michiel van Meeteren

Abstract. Social networks and social capital are generally considered to be important variables in explaining the diffusion of behavior. However, it is contested whether the actual social connections, cultural discourse, or individual preferences determine this diffusion. Using discrete choice analysis applied to longitudinal Twitter data, we are able to distinguish between social network influence on one hand and cultural discourse and individual preferences on the other hand. In addition, we present a method using freely available software to estimate the size of the error due to unobserved correlated effects. We show that even in a seemingly saturated model, the log likelihood can increase dramatically by accounting for unobserved correlated effects. Furthermore the estimated coefficients in an uncorrected model can be significantly biased beyond standard error margins.

1 Introduction

With the onset of ubiquitous social media technology, people leave numerous traces of their social behavior in – often publicly available – data sets. In this paper we look at a virtual community of independent ("Indie") software developers for the Macintosh and iPhone that use the social networking site Twitter. Using Twitter's API, we collect longitudinal data on network connections among the Indie developers and their friends and followers (approximately 15,000 nodes) and their use of Twitter client software over a period of five weeks (more than 600,000 "tweets"). We use this dynamic data on the network and user behavior to analyze the diffusion of Twitter client software.

Elenna R. Dugundji · Michiel van Meeteren
Universiteit van Amsterdam, Amsterdam, Netherlands
e-mail: {e.r.dugundji,michielvanmeeteren}@gmail.com

Ate Poorthuis
University of Kentucky, Lexington, KY, USA
e-mail: atepoorthuis@gmail.com

G. Ghoshal et al. (Eds.): *Complex Networks IV*, SCI 476, pp. 81–92.
DOI: 10.1007/978-3-642-36844-8_8

Within the Indie community, four prominent software developers have developed Twitter clients (Tweetie, Twitterrific, Twittelator, Birdfeed) that compete for adoption within the community. Apart from these Indie Twitter clients, members of the virtual community can choose from a range of clients that are developed outside of the Indie community (for example, Tweetdeck, Twitterfon) as well as the standard Web interface provided by Twitter. Previous qualitative ethnographic evidence for our case study indicates that social networks and social capital are considered to be important factors in explaining the adoption and diffusion of behavior. [1] Using discrete choice analysis applied to longitudinal panel data, we are able to quantitatively test for the relative importance of global cultural discourse, taste-maker influence and other contextual effects, node level behavioral characteristics, socio-centric network measures and ego-centric network measures, individual preferences and social network contagion, in users' decisions of what client software they choose to interface to Twitter.

Importantly, we furthermore demonstrate a method using readily available software to estimate the size of the error due to unobserved correlated effects in users' choices. This is critical to test for in any application of multinomial logistic regression where social influence variables and/or other network measures are used as explanatory variables, since their use poses a classic case of endogeneity. We show that even in a seemingly saturated model, the log likelihood of the model fit can increase significantly by accounting for unobserved correlated effects. Furthermore the estimated coefficients in the uncorrected model can be significantly biased beyond standard error margins. Failing to account for correlated effects can yield misleading market share predictions for users' preferences for Twitter clients.

The paper is organized as follows. First a brief review of literature is presented describing what the paper brings to an existing stream of behavioral modeling research. Next the understanding of the context of the case study and the insights from the available data lead us to define nine sets of different kinds of social and individual explanatory variables to explore in our model, with different functional forms. Estimation results are summarized. Finally, directions for future research efforts are outlined.

2 Discrete Choice with Social Interactions

Discrete choice analysis allows prediction based on computed individual choice probabilities for heterogeneous agents' evaluation of alternatives. In this section we first review the classic multinomial logit model and show how social interactions can be included in the model. Then we discuss an important econometric issue that arises in the empirical estimation of discrete choice models with social interactions, and present an approach to capture unobserved correlated effects. Finally we explain how the model can be approximated via simulation.

2.1 Multinomial Logit Model

In accordance with notation and convention in Ben-Akiva and Lerman [2], the multinomial logit model is specified as follows. Assume a sample of N decision-making entities indexed $(1,...,n,...,N)$ each faced with a choice among J_n alternatives indexed $(1,...,j,..., J_n)$ in subset C_n of some universal choice set C.

The choice alternatives are assumed to be mutually exclusive (a choice for one alternative excludes the simultaneous choice for another alternative, that is, an agent cannot choose two alternatives at the same moment in time) and collectively exhaustive within C_n (an agent must make a choice for one of the options in the agent's choice set). In general the composite choice set C_n will vary in size and content across agents: not all elemental alternatives in the universal choice set may be available to all agents. For simplicity in this paper however, we will assume that the choices are available to all agents.

Let $U_{in} = V_{in} + \varepsilon_{in}$ be the utility that a given decision-making entity n is presumed to associate with a particular alternative i in its choice set C_n, where V_{in} is the deterministic (to the modeler) or so-called "systematic" utility and ε_{in} is an error term. Then, under the assumption of independent and identically Gumbel distributed disturbances ε_{in}, the probability that the individual decision-making entity n chooses alternative i within the choice set C_n is given by:

$$\begin{aligned} P_{in} &\equiv P_n(i \mid C_n) = \Pr\left(V_{in} + \varepsilon_{in} \geq V_{jn} + \varepsilon_{jn}, \forall j \in C_n\right) \\ &= \Pr\left[V_{in} + \varepsilon_{in} \geq \max_{j \in C_n}\left(V_{jn} + \varepsilon_{jn}\right)\right] = \frac{e^{\mu V_{in}}}{\sum_{\forall j \in C_n} e^{\mu V_{jn}}} \end{aligned} \quad (1)$$

where μ is a strictly positive scale parameter which is typically normalized to 1 in the multinomial logit model.

The systematic utility is commonly assumed to be defined by a linear-in-parameters function of observable characteristics $\mathbf{S}_n$ of the decision-making entity and observable attributes $\mathbf{z}_{in}$ of the choice alternative for a given decision-making entity:

$$V_{in} = h_i + V\left(\mathbf{S}_n, \mathbf{z}_{in}\right) = h_i + \boldsymbol{\gamma}_i{}'\mathbf{S}_n + \boldsymbol{\zeta}_i{}'\mathbf{z}_{in} \quad (2)$$

The term h_i is a so-called "alternative specific constant" (ASC), as good practice to explicitly account for any underlying bias for one alternative over another alternative. In other words, h_i reflects the mean of ε_{jn} - ε_{in}, that is, the difference in the utility of alternative i from that of j when all else is equal. Since it is the difference that is relevant, for a general multinomial case with J alternatives we can define a set of at most $J - 1$ alternative specific constants.

The terms $\boldsymbol{\gamma}_i = [\gamma_{i1}, \gamma_{i2}, \ldots]'$ and $\boldsymbol{\zeta}_i = [\zeta_{i1}, \zeta_{i2}, \ldots]'$ are vectors of unknown utility parameters respectively corresponding to the relevant observable agent characteristics $\mathbf{S}_n$, and observable agent-specific attributes $\mathbf{z}_{in}$ of the choice alternatives. In general the utility parameters may take alternative specific values, however when there is no variation of the agent characteristics $\mathbf{S}_n$ across the choice alternatives, we can define a set of at most $J - 1$ vectors of alternative specific coefficients for the case of the γ_i .

2.2 Social Interactions

An outstanding challenge in discrete choice analysis is the treatment of the interdependence of various decision-makers' choices [3,4]. Brock and Durlauf [5] introduce social interactions in multinomial discrete choice models by allowing a given agent's choice for a particular alternative to be dependent on the overall share of decision makers who choose that alternative. If the coefficient on this interaction variable is close to zero and not important relative to other contributions to the utility, then the distribution of decision-makers' choices will not effectively change over time in relation to other decision-makers' choices. However, if the coefficient on this interaction variable is positive and dominant enough relative to other contributions to utility, there may arise a runaway situation over time as all decision-makers flock to one particularly attractive choice alternative. In short, the specification captures social feedback between decision-makers that can potentially be reinforcing over the course of time. In diverse literature this is referred to as a social multiplier, a cascade, a bandwagon effect, imitation, contagion, herd behavior, etc. [6]

We introduce a social feedback effect among agents by allowing the systematic utility V_{in} to be a linear-in-parameter β first-order function of the proportion x_{in} of a given decision-maker's reference entities who have made this choice. Our model differs from the Brock and Durlauf model in that we consider non-global interactions. Agents see different proportions, depending on who their particular reference entities are. Additionally, we also consider various socio-centric and ego-centric network measures and other explanatory variables as contributions to the utility.

2.3 Endogeneity

An econometric issue that arises in empirical estimation of social interactions in discrete choice models using standard multinomial logistic regression however, is that the error terms are assumed to be identically and independently distributed across decision-makers. It is not obvious that this is in fact a valid assumption when we are specifically considering interdependence between decision-makers' choices. We might reason that if there is a systematic dependence of each decision-maker's choice on an explanatory variable that captures the aggregate choices of other decision-makers who are in some way related to that decision-maker, then there might be an analogous dependence in the error structure. Otherwise said, the same unobserved effects might be likely to influence the choice made by a given decision-maker as well as the choices made by those in the decision-maker's reference group, which is a classic case of endogeneity. The results and coefficients of such a model are likely to be biased. To try to separate out effects, it is therefore first and foremost critically important to begin with an as well-specified model as possible, making use of relevant available explanatory variables. [7]

Dugundji and Walker [8] illustrate issues in the empirical estimation of a discrete choice model with network interdependencies using mixed generalized extreme value model structures with pseudo-panel data. Several modeling strategies are presented to highlight hypothesized interaction effects. In absence of true panel data on interaction between identifiable decision-makers, they use a priori beliefs about the social and spatial dimension of interactions to formulate the connectivity of the network and use socioeconomic data for each respondent as well as the geographic location of each respondent's residence to define aggregate interactions by grouping agents into geographic neighborhoods and into socioeconomic groups where the influence is assumed to be more likely. Technically, however, interactions between identifiable decision-makers may also be modeled using the approach described given the availability of suitable data.

In our empirical case study on adoption of Twitter clients, we do indeed have available data on which identifiable agents (Twitter users) plausibly influence other identifiable agents' choices, and furthermore we have longitudinal panel data observing repeated choices by agents over time. In this paper with such rich data, we continue this exploration of issues in the empirical estimation of discrete choice models with social interactions. Since our data is fairly large -more than 10,000 agents- we argue that the effect of unobserved correlated effects as perceived by any given agent is normally distributed, but is the same for that agent over the fairly short time period of the data collection. This simplified assumption allows us to specifically control for correlations in the error structure, through the use of mixed multinomial logit models with panel effects. [9].

2.4 Capturing Unobserved Correlated Effects

Suppose each agent n makes a sequence of choices at a number of points in time indexed $(1,\ldots,t,\ldots,T_n)$. For our case study, we will consider a general case where the number T_n of decision-making moments per agent varies across agents. We introduce an additive, normally-distributed agent-specific error term for each alternative i as follows:

$$U_{int} = V_{int} + \varepsilon_{int} + \sigma_i \xi_{in} \ ; \ \xi_n \sim N(0, I) \tag{3}$$

Conditional on ξ_n, the probability that agent n makes a particular sequence of choices over time $(i_1,\ldots,i_{Tn})$ is given by the product of the probabilities for agent n making each individual choice i_t:

$$P_n(i_1,\ldots,i_{T_n} \mid \xi_n) = \prod_{\forall t \in T_n} \frac{e^{\mu(V_{int}+\sigma_i \xi_{in})}}{\sum_{\forall j \in C_n} e^{\mu(V_{jnt}+\sigma_j \xi_{jn})}} \quad (4)$$

The unconditional user choice probability is the integral of this product over all values of ξ_n

$$P_n(i_1,\ldots,i_{T_n}) = \int_{\xi_n} \prod_{\forall t \in T_n} \frac{e^{\mu(V_{int}+\sigma_i \xi_{in})}}{\sum_{\forall j \in C_n} e^{\mu(V_{jnt}+\sigma_j \xi_{jn})}} N(0,I)d\xi_n \quad (5)$$

2.5 Econometric Estimation with Simulation

The unconditional choice probability is approximated through simulation for any given value of ξ_{in} as follows:

1. Draw a vector of values of ξ_n from $N(0, I)$ for each alternative in the choice set C_n, and label this ξ_n^r with the superscript $r = 1$ referring to the first draw
2. Calculate the conditional user choice probability for the particular sequence of choices made by agent n with this draw
3. Repeat steps 1 & 2 for R total number of draws and average the results

If the estimated coefficients σ_i can be shown to be statistically insignificant, we assume that the hypothesized endogeneity has negligible effect.

3 Modeling the Effects

This paper studies the diffusion of Twitter clients within the Indie community. Based on earlier research [10] we were able to determine a community of Indie developers that are actively using Twitter, using a mixed method community detection approach. For this community we use Twitter's publicly available API to gather data on network connections and actual messages sent. For 39 days, from 9 August until 16 September 2009, we harvested tweets and network connections on a daily basis for each of the nodes in the community.

Based on a review of the case study and the data [11], we expect client choice to be influenced by a number of distinct dimensions. Generally, social networks or social capital are considered to be important variables in explaining the adoption and diffusion of behavior. However, it is debatable to what extent the actual social connections, the global cultural discourse, and individual preferences influence this adoption and diffusion. Through our modeling of the effects, we can try to test the different hypotheses.

We distinguish between four Indie clients (Tweetie, Twitterific, Birdfeed and Twittelator), two popular non-Indie clients (Twitterfon and Tweetdeck) and the default Twitter web interface ("Web"). In addition, we employ a choice alternative, "Other" that serves as a baseline reference for the modeling. The "Other" category is highly heterogeneous and consists of more than 3500 clients that have relatively small market share (< 1%). On the basis of data we proceed to construct the nine sets of different kinds of social and individual explanatory variables to explore in our model.

3.1 Temporal Effects: Individual Preferences

We operationalize individual preference by constructing an alternative-specific relative individual cumulative lag variable. For each tweet, we count how often the sending user has been using each client in the seven days prior to sending the tweet resulting in an absolute cumulative lag variable. For each client, we then convert this absolute frequency to a relative cumulative lag variable indicating that client's use relative to how often that user has been using other Twitter clients in the past seven days. This individual preference variable shows how "sticky" a particular client has been for a user in the past seven days. This individual past behavior is likely to be a predictor of client choice for the next tweet, capturing complex UI preferences which we as researchers were not able to measure directly.

3.2 Temporal Effects: Social Network Contagion

To operationalize network influence we use the absolute cumulative lag variable as a basis. For each tweet, we count how often all users that the sender of that specific tweet is following use each client in the seven days prior to sending that the tweet. We convert the absolute frequency to an alternative specific relative network influence variable that indicates how often each client has been used relative to all other clients by all users that the sender of the tweet is following (ie. receiving information from). This can entail specific mentions of a client in a tweet but also more implicit or tacit knowledge about which client is popular or deemed useful within that user's social network. We argue that this usage by "friends" might influence client choice by either specific mentions of a client in Tweets or by the effect of tacit knowledge encoded within a user's social network.

3.3 Contextual Effects: Taste Maker Influence

We continue with exploring the contextual effect of whether or not a user in the community is connected to professional independent tech blogger John Gruber. Since Gruber promotes different clients to different extents [11], we are interested to see if the clients he promotes most favorably are used more often by the users connected to him. We operationalize this dummy variable in two different ways: if a user "follows" Gruber (ie. user receives tweets from Gruber); and if there is a reciprocal link with Gruber.

3.4 Contextual Effects: Developer Influence

Next, we are interested in the contextual effect of whether or not a user in the community is connected to a Twitter client developer [11] as follows: Clients developed by "Indies": Tweetie (Loren Brichter, Atebits); Twitterrific (Craig Hockenberry, Iconfactory); Twittelator (Andrew Stone, Stone Design); Birdfeed (Buzz Andersen, SciFi HiFi); Clients developed by others: TweetDeck (Iain Dodsworth, TweetDeck); TwitterFon (Kazuho Okui, Naan Studio). We operationalize each of these dummy variables in two different ways: if a user "follows" the developer (ie. user receives tweets from the developer); and if there is a reciprocal link with the developer (ie. the link with the developer is especially strong).

3.5 Behavioral Characteristics: Power Users

Since the Twitter clients have very different features, we might expect users who tweet a lot to prefer different kinds of clients than users who tweet less frequently. We operationalize this variable in four different ways: number of tweets sent by a user during observation period; "status count" (total tweets sent by a user during their entire history); number of tweets sent by a user prior to observation period (ie. giving emphasis of how active the user was in the past and how long the user has been using Twitter); and finally, the ratio of tweets sent by a user during observation period to total tweets sent during their entire history.

3.6 Network Measures: Central Users

As per our review of the importance of social media networks for "echo-chamber" marketing, we are interested in whether a user's position in the community affects their client choice. We compute five classic network centrality measures [12, 13]: in-degree centrality (the number of a user's "friends" in sample, ie. from whom tweets are received); out-degree centrality (the number of a user's "followers" in sample, ie. to whom tweets are sent); closeness centrality (sum of distances from a user to all other users, giving an indication of the expected time until arrival for information that might be flowing through the network); betweenness centrality (how often a user lies along the shortest path between two other users, giving an indication of access to diversity of information); and finally, eigenvector centrality (measures if a user is connected to many users who are themselves well connected, identifying users in centers of cliques).

3.7 Network Measures: Extended User In-Degree

In order to test the relative importance of the exposure to information flowing through the wider Twitter universe outside of the Indie community, we explore three extra network measure variables: the total number of a user's "friends" in the entire Twitter universe, ie. from whom a given user in principle receives tweets; the number of users outside the community from whom a given user in principle receives tweets; and finally, the ratio of users inside sample from whom a given user receives tweets to their total "friends" in the Twitter universe.

3.8 Network Measures: Extended User Out-Degree

Similarly, in order to test the relative opportunity to influence other users in the wider Twitter universe outside of the Indie community, we explore three extra network measure variables: the total number of a user's "followers" in the entire Twitter universe, ie. to whom a given user in principle sends tweets; the number of users outside the community to whom a given user in principle sends tweets; and finally, the ratio of users inside sample to whom a given user sends tweets to their total "followers" in the Twitter universe.

3.9 Global Influence

The cultural discourse on what is popular within the entire Indie community is operationalized by a set of alternative specific constants (ASC). Amongst things such as price and the impact of media exposure, we argue that this effectively captures global influence. It indicates the popularity of an alternative relative to all other alternatives during the entire sample period, after controlling for all other effects.

4 Results

All models are estimated using the freely available optimization toolkit Biogeme (http://biogeme.epfl.ch) developed by Bierlaire [14]. We begin by estimating a baseline multinomial logit model with alternative specific constants only, representing global bias. The log likelihood, number of estimated parameters and adjusted rho-squared are given in the first line of Table 1.

Next we test one-by-one each of the explanatory variables defined in Section 3.1-3.8. In cases where the variables are continuous (ie. for all cases except for the dummy variables in section 3.3 and 3.4), we also test linear, quadratic and square root forms of these variables. Based on log likelihood tests compared to the baseline model and *t*-tests on the estimated coefficients [2], we identify the best fitting variables per category. For example, the dummies defined as "follows Gruber" and "follows developer" are more significant than their respective forms "reciprocal link with Gruber" and "reciprocal link with developer"; the most significant centrality measures are closeness and square root of eigenvector centrality, etc. The interested reader is referred to [11] for details and interpretation.

Table 1 Log likelihood tests for incremental model specifications

Nr	Log Likelihood	Est. Par.	Rho Sq.	$-2[L_R - L_U]$	$\chi^2(0.1)$	p-Value
1	-968350.6	7	0.262	- -	- -	- -
2	-954368.7	14	0.272	27964	18.5	0.000
3	-568721.3	21	0.566	771295	18.5	0.000
4	-567945.3	28	0.567	1552	18.5	0.000
5	-566798.7	34	0.568	2293	16.8	0.000
6	-562010.9	41	0.571	9576	18.5	0.000
7	-561154.7	48	0.572	1712	18.5	0.000
8	-560664.6	55	0.572	980	18.5	0.000
9	-559662.6	62	0.573	2004	18.5	0.000
10	-559546.0	69	0.573	233	18.5	0.000
11	-452048.1	76	0.655	214996	18.5	0.000

1: Baseline model with alternative-specific constants only; 2: + Social network contagion (sq root); 3: + Lagged individual preferences (sq root); 4: + Follows Gruber; 5: + Follows developer; 6: + Frequency tweets during observation period (sq root); 7: + Eigenvector centrality (sq root); 8: + Closeness centrality; 9: + Ratio in-degree to total friends in Twitter (sq root); 10: + Ratio out-degree to total followers in Twitter (sq root); 11: + Estimated user-specific error component (R = 500 draws)

Having determined the best fitting variables and their respective functional forms, we then add the variables incrementally to the model, testing the improvement in log likelihood at each step. This is important to do, since variables that may have been significant when included in the model specification on their own, might no longer be significant when included together due to significance being shared between variables. The results are reported in lines 2-10 of Table 1. Each successive specification adds seven new parameters to the model (with the exception of "follows developer" where there are six since the Web alternative does not have a third party developer), as our data is rich and extensive enough to support alternative-specific definitions of the variables. In our case study, each new set of variables significantly improves the log likelihood (p-value of 0.000).

Finally, we include the normally-distributed user-specific error terms as in Section 2.4. We test the robustness of results using three different optimization algorithms for the maximization of the log likelihood, each with ten different random seeds for generating the draws. We use the estimated coefficients from the model in line 10 of Table 1 as a starting point for these 30 estimation runs with 50 draws, and then use the results with 50 draws in turn as the starting point for another 30 estimation runs with 200 draws, etc., for increasing number of draws, until the results stabilize across the random seeds for the three different optimization algorithms. Accounting for the unobserved correlated effects gave a dramatic jump in log likelihood as seen in line 11 of Table 1. The estimated coefficients in the final model in line 11 were also significantly different beyond standard error margins for 63 of 69 variables in the model in line 10. [11] Failing to account for unobserved correlated effects can thus yield misleading market share predictions for users' preferences for Twitter clients.

5 Conclusions and Recommendations

A prominent approach to studying the dynamics of networks and behavior stems from a growing stream of research on stochastic actor-based models. See Snijders, van de Bunt, and Steglich [15] for a tutorial. With the large data in our case study however, these established methods are not tractable. The alternative approach we discuss in this paper allows us to apply other freely available, open source, existing software for the estimation of the models. In so doing, we hope to stimulate researchers and practitioners to adopt these techniques when using large data sets of more than 1000 nodes due to the relatively lower entry barrier than could be the case if dedicated code would need to be written or if expensive software would need to be purchased. An interesting direction for further discrete choice research on diffusion in large networks may be combining the approach of Aral, Muchnik and Sundararajan [16] for distinguishing causal effects using propensity score matched sample estimation [17] in dynamic networked settings, with the present work accounting for unobserved correlated effects.

References

1. van Meeteren, M.: Indie fever: The genesis, culture and economy of a community of independent software developers on the Macintosh OS X platform. A Sofa Publication (2008), http://www.madebysofa.com/indiefever
2. Ben-Akiva, M., Lerman, S.R.: Discrete Choice Analysis: Theory and Application to Travel Demand. MIT Press, Cambridge (1985)
3. McFadden, D.: Economic choices. American Economic Review 91(3), 351–378 (2001)
4. McFadden, D.: Sociality, rationality, and the ecology of choice. In: Hess, S., Daly, A. (eds.) Choice Modelling: The State-of-the-Art and the State-of-Practice. Emerald Group Publ. Ltd., Bingley (2010)
5. Brock, W.A., Durlauf, S.N.: A multinomial choice model of neighborhood effects. American Economic Review 92(2), 298–303 (2002)
6. Manski, C.F.: Identification Problems in the Social Sciences. Harvard Univ. Press, Cambridge (1995)
7. Dugundji, E.R.: Socio-Dynamic Discrete Choice: Theory and Application. Ph.D. manuscript. Universiteit van Amsterdam, Netherlands (2012)

8. Dugundji, E.R., Walker, J.L.: Discrete choice with social and spatial network interdependencies: An empirical example using mixed generalized extreme value models with field and panel effects. Transportation Research Record 1921, 70–78 (2005)
9. Train, K.E.: Discrete Choice Methods with Simulation, 2nd edn. Cambridge University Press, New York (2009)
10. Meeteren, M., Poorthuis, A., Dugundji, E.R.: Mapping communities in large virtual social networks. In: IEEE International Workshop on Business Applications of Social Network Analysis, Bangalore, India (2010), doi:10.1109/BASNA.2010.5730297
11. Dugundji, E.R., Poorthuis, A., van Meeteren, M.: Capturing unobserved correlated effects in diffusion in large virtual networks: Distinguishing individual preferences, social connections and cultural discourse influence on the adoption of Twitter clients (2012) (unpublished)
12. Freeman, L.C.: Centrality in social networks. Social Networks 1(3), 215–239 (1978)
13. Bonacich, P.: Power and centrality: A family of measures. Am. J. Soc. 92(5), 1170–1182 (1987)
14. Bierlaire, M.: Biogeme: A free package for the estimation of discrete choice models. In: Proceedings of the 3rd Swiss Transportation Research Conference, Ascona, Switzerland (2003)
15. Snijders, T.A.B., van de Bunt, G.G., Steglich, C.E.G.: Introduction to stochastic actor-based models for network dynamics. Social Networks 32(1), 44–60 (2010)
16. Aral, S., Muchnik, L., Sundararajan, A.: Distinguishing influence-based contagion from homophily-driven diffusion in dynamic networks. Proc. Nat'l Acad. Sci. 106(51), 21544–21549 (2009)
17. Hill, S., Provost, F., Volinsky, C.: Network-based marketing: Identifying the likely adopters via consumer networks. Statistical Science 21, 256–276 (2006)

Oscillator Synchronization in Complex Networks with Non-uniform Time Delays

Jens Wilting and Tim S. Evans

Abstract. We investigate a population of limit-cycle Kuramoto oscillators coupled in a complex network topology with coupling delays introduced by finite signal propagation speed and embedding in a ring. By numerical simulation we find that in complete graphs velocity waves arise that were not observed before and analytically not understood. In regular rings and small-world networks frequency synchronization occurs with a large variety of phase patterns. While all these patterns are nearly equally probable in regular rings, small-world topology sometimes prefers one pattern to form for a large number of initial conditions. We propose implications of this in the context of the temporal coding hypothesis for information processing in the brain and suggest future analysis to conclude the work presented here.

Introduction

The study of coupled oscillators and their synchronization behavior [3] has been a field of study for a long time with widely spread applications like flashing fireflies, cardiac pacemaker cells or epileptic seizures [15, 17]. Furthermore the interplay of oscillators is also subject of interest in the context of the temporal coding hypothesis [5, 7] which suggests that the information emitted by a population of neurons

Jens Wilting · Tim S. Evans
Networks and Complexity Programme, Imperial College London,
London SW7 2AZ, United Kingdom

Jens Wilting
Dept. of Epileptolgy, University of Bonn, Sigmund-Freud-Strasse 25,
D-53105 Bonn, Germany
e-mail: jwilting@uni-bonn.de

Tim S. Evans
Theoretical Physics, Blackett Laboratory, Imperial College London,
London SW7 2AZ, United Kingdom
e-mail: t.evans@imperial.ac.uk

G. Ghoshal et al. (Eds.): *Complex Networks IV*, SCI 476, pp. 93–100.
DOI: 10.1007/978-3-642-36844-8_9

is encoded by the timing of firing action potentials, and therefore depends on the coupling within this population.

A simple model for synchronization of limit-cycle phase oscillators is given by the Kuramoto model [1, 14]. If the coupling is constraint by an underlying network topology, i.e. two oscillators or nodes are coupled if there is a link between them in the network, it has been shown that the network structure itself influences the synchronization of oscillators [2, 4, 8, 9, 10, 16, 19] and can either favor or hinder coherence.

In most applications the coupling cannot be regarded as instantaneous but finite signal propagation speeds impose delays, e.g. the speed of sound or axonal signal speeds in nerve cells. In the case of identical delays for each pair of coupled oscillators this leads to vastly enhanced synchronization phenomenology [6, 22].

However often one would not expect the delays to be identical for all oscillators. Then in the most general form the Kuramoto model for N limit-cycle phase oscillators is extended to the form

$$\dot{\theta}_i(t) = \omega_i - \frac{1}{k_i}\sum_{j=1}^{N} K_{ij}\sin\left(\theta_i(t) - \theta_j(t-\tau_{ij})\right). \tag{1}$$

with the degree k_i of node i to incorporate non-uniform time delays τ_{ij} and coupling strengths K_{ij} between oscillators i and j with phases θ_i and θ_j respectively. The underlying network topology with adjacency matrix A_{ij} is encoded in the coupling strengths such that $K_{ij} = K \times A_{ij}$.

In the case of delays induced by signal propagation over a distance d_{ij} with finite signal speed v the delays are given as $\tau_{ij} = d_{ij}/v$. Since networks do not have an inherent metric the oscillator population is embedded on a one-dimensional unit circle [23] with oscillator positions ϕ_i. The distance between two oscillators is then given by the minimum distance on this circle,

$$d_{ij} = \min\left\{|\phi_i - \phi_j|, 2\pi - |\phi_i - \phi_j|\right\}, \tag{2}$$

which behaves like a well-defined metric. The oscillators are supposed to be evenly spread around the ring, thus the positions are given by

$$\phi_i = \frac{2\pi}{N}\cdot(i-1). \tag{3}$$

In this setup with global coupling $K_{ij} = K$ for all i, j depending on the propagation speed full synchronization and waves traveling around the ring arise [23], a feature that was also observed in a two-dimensional metric space [11] and if random graphs are chosen for the coupling topology [12, 13].

Since analytical results can only be obtained for a limited number of cases, one of them being presented for global coupling in [23], we make wide use of numerical simulations. A fourth order Runge Kutta scheme, extended by a backwards search in time to incorporate the time delays, is used for integration. Since the delays require an evaluation at $t < 0$ the system is integrated from $t = 0$ to $t = -\max\left\{\tau_{ij}\right\}$ with a backwards Euler method and free oscillator evolution $\dot{\theta} = \omega_i$, i.e. the coupling is supposed to set in at $t = 0$ and is in fact given by $K_{ij} \times H(t)$ with the Heavyside step function $H(t)$.

In accordance with previous work on this particular model we choose identical frequencies $= \omega_i = 1.0$ and $K = 1.0$ while the signal speed is drawn from the interval $v \in [0.15, 1.35]$. The system is simulated from $t = -\max\{\tau_{ij}\}$ to $t = 200$ with a step size of 0.0025. Every simulation is repeated twenty times with different sets of initial conditions $\theta_i(0)$ drawn from a uniform random distribution in $[0, 2\pi)$ to reveal potentially critical dependence on initial conditions. Different types of networks with 100 nodes each are chosen for the underlying coupling topology.

Global Coupling

With all-to-all coupling, $K_{ij} = K$ for all i, j, we reproduce the results of [23] and find that for all initial conditions the system converges to full phase coherence or phase waves traveling around the ring, whose winding number m depends on the size of the delays. It can be extracted from the simulation by summation over all phase differences between pairs of adjacent oscillators, where the ambiguity of 2π shifts is resolved by requiring the difference to be within $[-\pi, \pi]$:

$$m(t) = \frac{1}{2\pi} \sum_{i=1}^{N} f\left(\theta_i(t) - \theta_{i+1}(t)\right), \quad \text{where } f(x) = \begin{cases} x, & \text{if } |x| \leq \pi \\ x - 2\pi, & \text{if } x > \pi \\ x + 2\pi, & \text{if } x < -\pi. \end{cases} \tag{4}$$

Analysis reveals that the winding number defined in this way after initial fluctuations converges to a stable value. These fluctuations represent the transient from initial conditions to the observed waves: From randomly distributed initial conditions, yielding random winding number m, growing segments of the ring synchronize or form waves. These waves may have different directions so that typically the winding number goes down to a value close to 0 at some point, until two large segments with opposite waves form. When these segments gradually merge into one globally coherent pattern the winding number converges to its final value.

Figure 1 shows that the final winding number observed in the system increases with the delays, i.e. with $1/v$. This increase appears to be nearly linear in the investigated parameter range, but is quantized to integer values of the winding number. Furthermore the mapping of $1/v$ to $|m|$ is uniquely defined except near the transitions to higher $|m|$ where two different winding numbers can manifest dependent on the initial conditions. The one exception at $1/v \approx 6.7$ is due to the fact that for large delays the transient to a stable wave-mode is not completed when the simulation finishes at $t = 200$, so that at this time two large segments with opposite waves exist, canceling their phase differences to $|m| = 0$.

Zanette [23] observed that all oscillators, whose velocity is $\Omega_i = \dot{\theta}_i$, lock to one common frequency Ω so that the velocity dispersion $\sigma = \sqrt{\left\langle (\Omega_i - \langle \Omega \rangle)^2 \right\rangle}$ decays

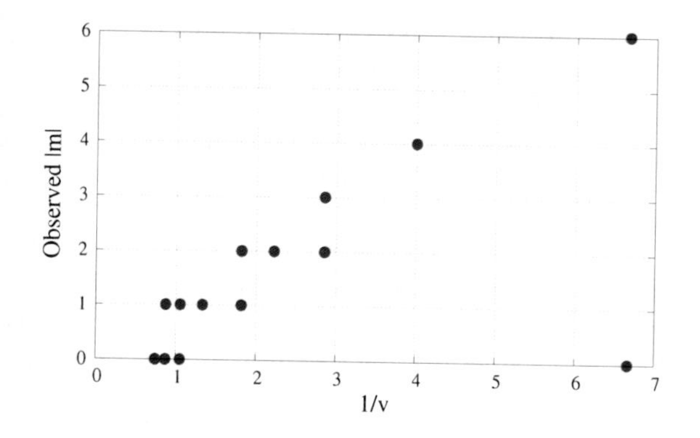

Fig. 1 Final state winding number observed for at least one set of initial conditions versus signal propagation speed. $|m|(v)$ is uniquely defined except near the transitions to higher $|m|$, where different initial conditions may lead to different winding numbers.

with a decay constant dependent on the size of the delays. We confirm that for high signal speed, i.e. small delays, this is indeed the case but for signal speed $v < 0.75$ find that the velocity dispersion converges to a finite limit (see Fig. 2). Closer analysis of the velocities shows that for each oscillator Ω_i is *not* a constant value but periodically oscillates around Ω. The phase of these oscillations is shifted by π for opposite oscillators, so that in fact also the instantaneous velocity of the oscillators forms a wave traveling around the ring. The peak-to-peak amplitude of this oscillation ranges from 0.05 up to 0.8 compared to an average velocity of $\Omega \approx 0.9$. So far we do not know why these velocity oscillations arise and why they were not seen before. We observe them consistently for all initial conditions and with a simulation step size varied over 2.5 decades, making it seem improbable that they are a mere artifact of the simulation. Furthermore the limit of the velocity dispersion depends only on v and is the same for all initial conditions if v is fixed. Attempts to understand these waves as a perturbations of the frequency-locked wave modes that were analytically understood [23] have not been successful so far and are restricted to assumptions that are valid in a very limited number of observed cases only, so that more effort will be needed to explain the observed phenomenon.

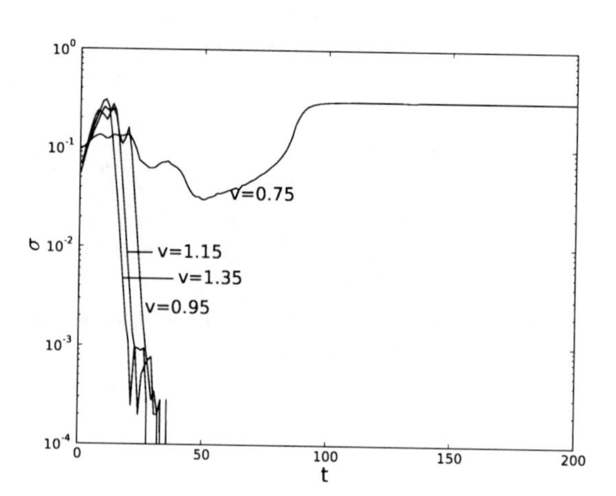

Fig. 2 The velocity dispersion σ decays exponentially for $v > 0.95$ but converges to a finite limit for smaller speeds. This is the result of each single oscillator velocity oscillating around an average value, so that in fact the velocity forms a wave traveling around the ring in addition to the known phase waves.

Regular Rings and Small-Worlds

Regular rings are inherently connected to the embedding in a circle. Therefore we examine networks in which each node is connected to its k nearest neighbors where k is even.

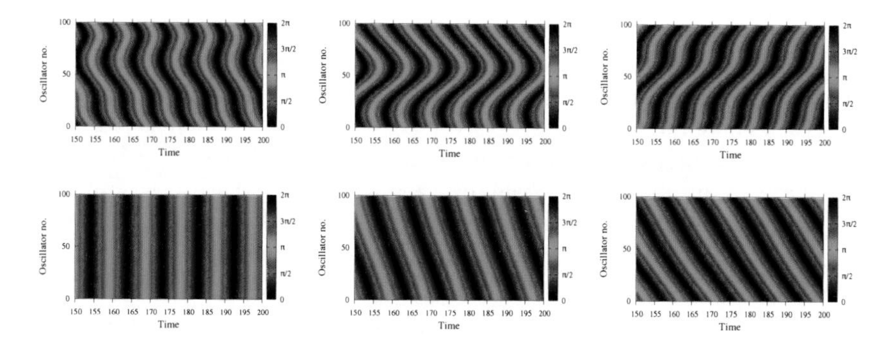

Fig. 3 Spatio-temporal phase representation for regular rings and $\nu = 0.35$. In all cases a steady state evolution with a stable propagating phase pattern emerges. First row: In a bidirectional ring with $\bar{k} = 2$ a different, possibly complicated, phase pattern emerges for each set of initial conditions. Third row: Only full synchronization and wave modes occur for $\bar{k} = 10$.

We observe that in all cases the oscillators synchronize in frequency after an initial transient and converge to a stable state where they lock to one phase pattern and propagate at a common frequency. However this phase pattern turns out to be different for every realization of initial conditions in a bidirectional ring with only nearest neighbor connections and average degree $\bar{k} = 2$. Furthermore more complex patterns than waves occur as is depicted in the first row of Fig. 3.

For an explanation of this difference from the globally connected case follow the analysis in [23] and assume that the oscillators lock to a common frequency Ω and evolve as $\theta_i(t) = \Omega t + \hat{\theta}_i$. Substituting this in the model equation (1) yields

$$\Omega = \omega - \frac{1}{k_i} \sum_{j=1}^{N} K_{ij} \sin\left(\Omega \tau_{ij} + \hat{\theta}_i - \hat{\theta}_j\right). \tag{5}$$

For global coupling $K_{ij} = K$ the requirement of the left hand side of this equation being independent of i restricts possible choices of the $\hat{\theta}_i$ and leads to wave modes [23]. In a ring however only two terms contribute to the sum and give the new condition

$$\Omega = \omega - \frac{K}{2} \left[\sin\left(\Omega \tau + \hat{\theta}_i - \hat{\theta}_{i-1}\right) + \sin\left(\Omega \tau + \hat{\theta}_i - \hat{\theta}_{i+1}\right)\right] \tag{6}$$

where $\tau = \tau_{i,i\pm 1}$. Define $\Delta_i = \hat{\theta}_i - \hat{\theta}_{i-1}$, then then a solution consistent with frequency synchronization must give the same result Ω for all i, i.e.

$$\sin(\Omega\tau + \Delta_i) + \sin(\Omega\tau - \Delta_{i+1}) = \text{const.} \tag{7}$$

By fixing two phases $\hat{\theta}_i, \hat{\theta}_{i+1}$ this equation fixes all other θ_j. Note however that only solutions for Δ_i are possible that sum up to integer multiples of 2π and must comply with the self-consistency equation (6) for Ω. Nevertheless this equation imposes less constraints on possible phase differences between oscillators, so that more phase patterns are consistent with locking to a common frequency Ω and can arise. This conforms to the observation that in regular rings with coupling to the next five neighbors on each side, where more terms contribute to the sum in equation (5) and impose more constraints, only wave modes and complete synchronization are observed (second row in Fig. 3) similar to the case of global coupling in [23].

From regular rings small-world networks are created [20] by rewiring links to a random new destination with probability p and thus adding shortcuts and disorder into the network. We inspect networks with average degrees 4 and 12 and find that the system behavior strongly depends on the rewiring probability p: In the limit of regular rings, $p < 0.05$, the behavior of rings is reproduced, i.e. wave modes manifest for higher degree $\bar{k} = 12$ while a large variety of stable phase configurations is observed for $\bar{k} = 4$. In contrast to the purely regular case there are phase slips that divide the pattern into separate coherent parts. These slips can be mapped to the few shortcuts in the system. Increasing the amount of disorder in the network for $p > 0.5$ the observations for random graphs [12, 13] are reproduced: The stable phase patterns show no obvious coherence for $\bar{k} = 4$ and are waves for $\bar{k} = 12$.

More connected graphs show an interesting behavior: While in both limits of p waves are observed, the range of intermediate $0.1 < p < 0.5$ produces a large variety of phase patterns as seen in regular rings. However there is an important difference, as some network realizations seem to highly favor one certain pattern to form for a large number of initial conditions.

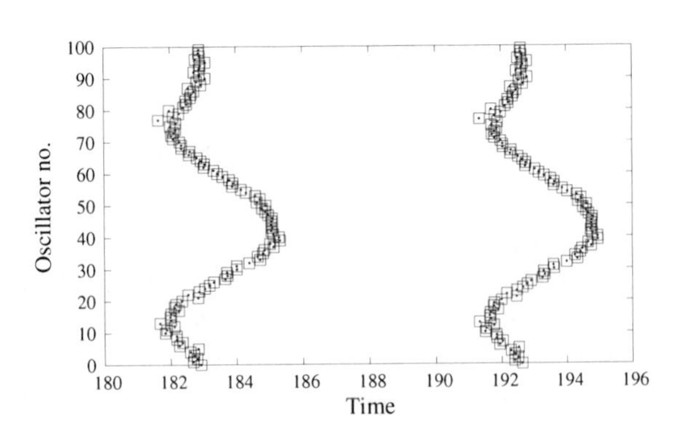

Fig. 4 Defining an oscillator to fire when passing a phase of 2π from below a firing sequence similar to those in [21] is observed in one special realization of small-world topology with $p = 0.3$ for 80% of initial conditions. Other small-world realizations may lead to other firing patterns with large basin of attraction or have no pattern produced by a significantly large percentage of initial conditions.

In the light of neuronal networks and the temporal coding hypothesis [5] it has been shown [18, 21] that rings of unidirectionally coupled oscillators can produce predefined temporal firing sequences by adjusting the coupling delays appropriately as can be done by synaptic plasticity in the brain. The results obtained in this work suggest that also special network topology can be used to generate one certain firing pattern with high probability. Defining an oscillator to fire when passing a phase of 2π from below, Fig. 4 depicts the periodically occurring firing sequence of an oscillator population that was observed for 80% of initial conditions. A large variety of other patterns with large basin of attraction was observed for other realizations of small-world networks. So far we have not fully understood the connection between network and resulting pattern, although there are hints for a connection between high closeness centrality of a section of the ring and late firing in the sequence of the corresponding oscillators and early firing for oscillator sections with lower closeness centrality which would indicate a connection to information spread within the network.

Conclusion

In summary we have investigated the synchronization of limit-cycle oscillators with distance dependent coupling delays introduced by finite signal propagation speed. For global coupling topology we mostly confirm prior results [13, 23] but find differences in the frequency synchronization, where we observe frequency waves traveling around the ring in addition to the phase waves.

The smaller number of coupling partners in (7) for regular rings allows more phase patterns to stably exist in the oscillator population consistent with frequency locking. While for regular rings the developing pattern was different for almost every set of initial conditions, some realizations of small-world connectivity with a small amount of disorder or shortcuts result in a large basin of attraction of one certain phase pattern, which is favored to form for a large number of initial conditions. This has implications for the creation of temporal firing patterns which seems possible through special small-world topology.

From this point many directions are open for future research like breaking the symmetry of evenly spread oscillator positions on the ring or inhomogeneous coupling constants.

Acknowledgements. We would like to thank Michael Gastner, Tiago Pereira, Jeldtoft Jensen and Gunnar Pruessner for their advice and help.

References

1. Acebrón, J.A., Bonilla, L.L., Pérez Vicente, C.J., et al.: The Kuramoto model: A simple paradigm for synchronization phenomena. Rev. Mod. Phys. (2005), doi:10.1103/RevModPhys.77.137

2. Arenas, A., Díaz-Guilera, A., Kurths, J., et al.: Synchronization in complex networks. Phys. Rep. (2008), doi:10.1016/j.physrep.2008.09.002
3. Boccaletti, S., Kurths, J., Osipov, G., et al.: The synchronization of chaotic systems. Phys. Rep. (2002), doi:10.1016/S0370-1573(02)00137-0
4. Brede, M.: Locals vs. global synchronization in networks of non-identical Kuramoto oscillators. Eur. Phys. J. B (2008), doi:10.1140/epjb/e2008-00126-9
5. Cessac, B., Paugam-Moisy, H., Viéville, T.: J. Physiol. Paris (2010), doi:10.1016/j.jphysparis.2009.11.002
6. Choi, M.Y., Kim, H.J., Kim, D., Hong, H.: Synchronization in a system of globally coupled oscillators with time delay. Phys. Rev. E (2000), doi:10.1103/PhysRevE.61.371
7. Foss, J., Longtin, A., Mensour, B., Milton, J.: Multistability and delayed recurrent loops. Phys. Rev. Lett. (1996), doi:10.1103/PhysRevLett.76.708
8. Gomez-Gardenes, J., Moreno, Y., Arenas, A.: Paths to Synchronization on Complex Networks. Phys. Rev. Lett. (2007), doi:10.1103/PhysRevLett.98.034101
9. Hong, H., Choi, M.Y., Kim, B.J.: Synchronization on small-world networks. Phys. Rev. E (2002), doi:10.1103/PhysRevE.65.026139
10. Ichinomiya, T.: Frequency synchronization in a random oscillator network. Phys. Rev. E (2004), doi:10.1103/PhysRevE.70.026116
11. Jeong, S.-O., Ko, T.-W., Moon, H.-T.: Time-Delayed Spatial Patterns in a Two-Dimensional Array of Coupled Oscillators. Phys. Rev. Lett. (2002), doi:10.1103/PhysRevLett.89.154104
12. Ko, T.-W., Ermentrout, G.B.: Effects of axonal time delay on synchronization and wave formation in sparsely coupled neuronal oscillators. Phys. Rev. E (2007), doi:10.1103/PhysRevE.76.056206
13. Ko, T.-W., Jeong, S.-O., Moon, H.-T.: Wave formation by time delays in randomly coupled oscillators. Phys. Rev. E. (2004), doi:10.1103/PhysRevE.69.056106
14. Kuramoto, Y.: Chemical Oscillations, Waves, and Turbulence. Springer, Berlin (1984)
15. Lehnertz, K., Bialonski, S., Horstmann, S.T., et al.: Synchronization phenomena in human epileptic brain networks. J. Neurosi. Meth. (2009), doi:10.1016/j.jneumeth.2009.05.015
16. Moreno, Y., Pacheco, A.F.: Synchronization of Kuramoto oscillators in scale-free networks. Europhys. Lett. (2004), doi:10.1209/epl/i2004-10238-x
17. Mormann, F., Lehnertz, K., David, P., Elger, C.E.: Mean phase coherence as a measure for phase synchronization and its application to the EEG of epilepsy patients. Physica D (2000), doi:10.1016/S0167-2789(00)00087-7
18. Popovych, O.V., Yanchuk, S., Tass, P.A.: Delay- and Coupling-Induced Firing Patterns in Oscillatory Neural Loops. Phys. Rev. Lett. (2011), doi:10.1103/PhysRevLett.107.228102
19. Strogatz, S.H., Mirollo, R.E.: Phase-locking and critical phenomena in lattices of coupled nonlinear oscillators with random intrinsic frequencies. Physica D (1988), doi:10.1016/0167-2789(88)90074-7
20. Watts, D.J., Strogatz, S.H.: Collective dynamics of 'small-world' networks. Nature (1998), doi:10.1038/30918
21. Yanchuk, S., Perlikowski, P., Popovych, O., et al.: Variability of spatio-temporal patterns in non-homogeneous rings of spiking neurons. Chaos (2011), doi:10.1063/1.3665200
22. Yeung, M.K., Strogatz, S.H.: Time Delay in the Kuramoto Model of Coupled Oscillators. Phys. Rev. Lett. (1999), doi:10.1103/PhysRevLett.82.648
23. Zanette, D.H.: Propagating structures in globally coupled systems with time delays. Phys. Rev. E (2000), doi:10.1103/PhysRevE.62.3167

Understanding History Through Networks: The Brazil Case Study

Hugo S. Barbosa-Filho, Fernando B. de Lima-Neto, and Ronaldo Menezes

Abstract. In the 19th century Alphonse the Lamartine (1790–1869), a French writer, said that *"History Teaches Everything, Including the Future."* This is a very accurate statement; the definition of what makes whole nations, what characterizes patriotism, relates to the history of that particular place. This understanding about the importance of History has always been recognized by thinkers, philosophers and writers. However, historical facts are often source of controversies and debates. In this paper we have applied Network Science techniques to analyze a Social Network of the History of Brazil to create a rank of Historical Characters. To carry out such analyses we first have built a dataset based on Wikipedia text bodies using Natural Language Processing.

1 Introduction

In the past decades, Network Science has been successfully applied in historical studies using a wide variety of data sources such as maps, documents and texts [1]. Given the own nature of the historical method (often based on historian's interpretation), historical facts are often controversial and a source of debates. A historical figure can be either a hero or a criminal, depending on the historian's ideology, culture and values. Historial Figures such as Joseph Stalin and Mao Tse Tung (to mention a few) are more than controversial [2]. The fact is that defining the relevance of

Hugo S. Barbosa-Filho · Ronaldo Menezes
BioComplex Laboratory, Computer Science Department, Florida Institute of Technology, 150 W University Blvd. Melbourne - FL, USA
e-mail: hbarbosafilh2011@my.fit.edu, rmenezes@fit.edu

Fernando B. de Lima-Neto
Polytechnique School of Pernambuco,
University of Pernambuco, Rua Benfica, 455, Recife-PE, Brazil
e-mail: fbln@ecomp.poli.br

G. Ghoshal et al. (Eds.): *Complex Networks IV*, SCI 476, pp. 101–108.
DOI: 10.1007/978-3-642-36844-8_10

a historical character can be misleading and severely subjective. A good approach to overcome this subjectivity (or at least reduce it) is to also perform quantitative analyses on historical data.

In this context, Network Science has proven to be a solid framework to assess and to ascribe importance within Social Sciences and particularly History [3]. In this paper we use Network Science to analyze a social network of individuals related to the history of Brazil extracted from Wikipedia entries.

2 Related Works

The knowledge the past plays a key role to the process of understanding the present and of being prepared for the future. Such fact has been recognized by many thinkers and writers. Mark Twain for instance argued that *"History never repeats itself, but it often rhymes."* In this sense, understanding the role of different characters in historical events have importance. It is this understanding that enables a more precise interpretation of present and future events.

As Borgatti *et al.* pointed out, over the past decade, techniques from Network Science received increasing attention from social scientists [3]. More precisely, historians have been applying network theories in investigations on a variety of subjects such as religious innovation and ancient civilizations [4, 5, 1]. However, building social networks from historical data is still challenging. Most of the issues relate to the lack of structure in many historical data sources such as documents, letters and newspapers [4]. Furthermore, the precise definition of what is a relationship within a text is still challenging and detection of semantic relations is an active problem in natural language processing and information retrieval [6].

From this scenario where information is plentiful but hard to be automatically extracted and processed the Semantic Web movement was created. This collaborative effort led by W3C aims to make the information available on the Web more structured so that it can be directly or indirectly processed by machines [7]. A worth-mentioning semantic web tool is DBPedia, whose objective is to extract structured content from Wikipedia, allowing users and machines to query relationships and properties from Wikipedia resources [8]. In the past few years, DBPedia has been increasingly used as source for the construction of datasets for network studies (e.g., Jazz musicians network [9], recommender systems [10]). Also, in a recent work, Historical events were successfully extracted from DBpedia [11].

However, semantic web tools are based on ontologies collaboratively built. In the case of Wikipedia, only 15% of *things* are described in DBpedia and only 7% are classified in a consistent ontology.[1] This means that most of the Wikipedia information is only contained in the article's bodies as unstructured or semi-structured text. In such scenarios, tools based on Natural Language Processing (NLP) have been of great help in extracting data and relationships [12, 13, 14].

[1] `http://dbpedia.org/About`

In this paper we have applied a well-known set of techniques for the first time in a (1) named-entity extraction to create a (2) Social Network of (3) Historical characters (in this case, from the History of Brazil) from (4) Wikipedia articles bodies.

3 A Social Network of Historical Characters

The Wikipedia Project is a crowdsourcing based encyclopedia with about 23.3 million articles and 17.7 million registered users (as of September 2012). Only a minority of them are active users (i.e. have contributed 5 times or more in one month).

To build our dataset we have created a tool to crawl, extract and process person names from Wikipedia articles that are somehow related to a given context, in this case, the brazilian history (see Figure 1 for an overall depiction to the process). The software was parametrized with domain-specific patterns aiming to bound the results to the subject. From a network perspective, *nodes* are the names extracted from the text. An *edge* between nodes A and B exists if B appears on A's Wikipedia article and vice-versa. The network we have built is *directed* and *unweighted*.

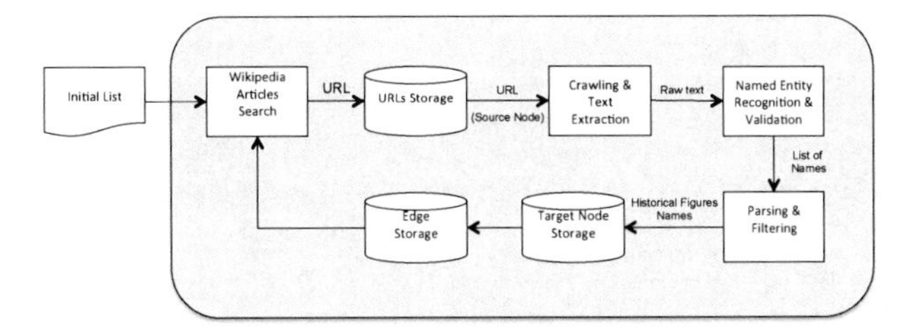

Fig. 1 General overview on the extraction process. From an "Initial List", the crawler loops through several steps and stops when the necessary set of URLs have been reached (user defined).

3.1 Building the Network

An initial list of 135 names was manually collected from a high-school History textbook [15]. For each entry in the list a query is made to Wikipedia search API [2] searching for entries whose title matches that name. The response is a JSON object with details for that entry. In order to find whether that article is of our interest, the crawler verifies if that page belongs to the *people* category. If it does, its URL is then returned. For the initial list of 135 names, 124 had its own article. This list of 124 pages will feed the crawler for its first iteration.

Starting the aforementioned list of 124 names, the Named-Entity Recognition and Validation (NERV) module extracts from the article body all names of people. Given that NERV tools should be 100% precise, we decided to use 3 NERV instances to

[2] `http://www.mediawiki.org/wiki/API:Search`

cross validate each possible name. The first one was an interface to AlchemyAPI, a Natural Language Processing API which performs several content analysis tasks such as named-entity extraction, concept tagging and sentiment analysis.[3] The second and the third were instances of Apache OpenNLP, a machine learning based toolkit for natural language processing maintained by The Apache Software Foundation.[4]. One of the instances works with Spanish and the other with Portuguese name finder models.[5]

A person name is considered valid if at least two NERV instances recognize it as a name. Then, each name is searched on Wikipedia to check if it has its own article. If it does, the article's text will be parsed according to a set of parameters (e.g. occupation and year of death) aiming to identify whether that page belongs to a historical character. The page is then inserted into the *pages* table, the name is inserted into the *characters* table. Finally, a connection is created from person being mentioned on the page to the subject of the page (who must be a person also).

Another somewhat common situation is when a person name is found in an article but that person does not have an article written on Wikipedia. In this case, the name is inserted into the *characters* table. In order to keep track of who was mentioned in whose article an entry is inserted into the *mentions* table.

4 Results

The focus of this section is on the network structure as well as on individual attributes. As pointed out in previous sessions, in this work we aim at answering the following questions: *(1)* is the network built from Wikipedia articles a social network?, *(2)* Who were the most relevant characters from the network perspective?, and *(3)* Is it possible to use network science to better understand the role that individuals actually played for the course of brazilian history?

To answer these questions we have first verified whether the network we have built is a real social network. As suggested by Watts and Strogatz,[16], to perform this verification we need to see if the network satisfies two basic conditions: it shows Small-World characteristics and the degree distribution fits well to a power-law distribution. We have compared the measures of our network with other well-known real networks. Table 1 shows that the network we have built can be characterized as a small-world network.

When we compare the average path length ℓ and the clustering coefficient C from the historical social network with other well-known real networks, we see that it has small-world properties. Moreover, the historical network shows a larger clustering coefficient $C = 0.12$ when compared to a random network with similar parameters ($C = 0.001$), suggesting that what we have is indeed a small-world network. When

[3] http://www.alchemyapi.com

[4] http://opennlp.apache.org

[5] The Portuguese name model was trained using a brazilian portuguese text corpus part of the Floresta Sintá(c)tica available at http://www.linguateca.pt/Floresta/

Table 1 Network metrics for Brazilian Historical Network compared to the Internet [17] and movie-actors networks [18] and the electric power grid of Southern California [16].

Variable Name	Historical Figures	Random	Movie-Actors	Internet	Power Grid
Nodes (n)	2,159	2,159	449,913	10,697	4,941
Edges (m)	4,922	4,922	25,516,482	31,992	6,594
Mean degree (z)	2.28	2.28	113.43	5.98	2.669
Out-degree Exp. of Power Law (λ)*	2.1	-	2.3	2.5	-
Avg. Clustering Coefficient (C)	0.12	0.001	0.2	0.035	0.08
Avg. Path Length (ℓ)	7.56	8.87	3.48	3.31	18.98

* The exponent corresponding to the in-degree power-law distribution fitness was omitted given that it does not follow a power-law distribution (more on that later on this section).

we plot the cumulative degree distribution in log-log scale (Figure 2) we can see that the degree distribution follows a power-law.

However, only the out-degree distribution follows a power law. This is due to the fact any node can be the source of a relationship while only those that have their own article can be a target. Thus, the much larger number of individuals who do not have their own page ($n = 1,526$) deformed the in-degree distribution. Figure 3 provides a visualization on the Social Network of Historical Characters (a) as well as a representation of a Self-Organizing Map (b) trained with nodes' measures (namely. *Degree*, *PageRank*, *Betweenness Centrality* and *Clustering Coefficient*).

In both PageRank and Eigenvector centrality analysis, we have observed that a large number of politicians appeared among the top ranked, even those that had held less relevant positions for a short period of time. Six out of 10 individuals with the highest PageRank and all top 10 personalities with highest eigenvector centralities

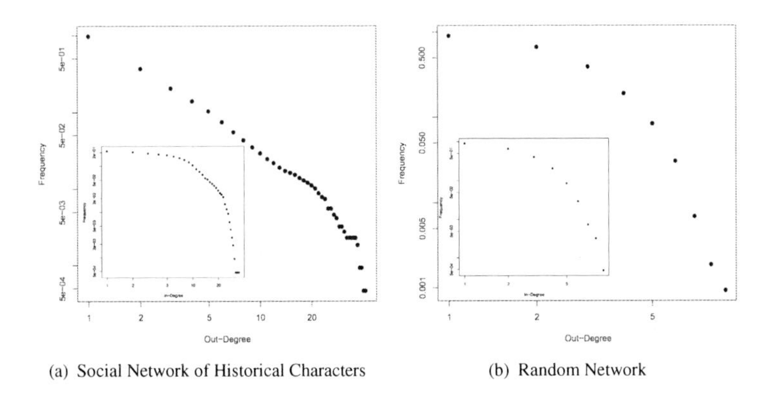

(a) Social Network of Historical Characters (b) Random Network

Fig. 2 Cumulative degree distribution. The main plot is the out-degree and in the inset the in-degree: (a) corresponds to the Social Network of Historical Characters. It is possible to see from the main plot in (a) that the out-degree distribution clearly follows a power-law with an exponential cut-off. On the other hand, the inset shows that the in-degree distribution rejects the power-law hypothesis. As a comparison, we show in (b) that a random network with similar parameters has different degree distribution and does not follow a power-law.

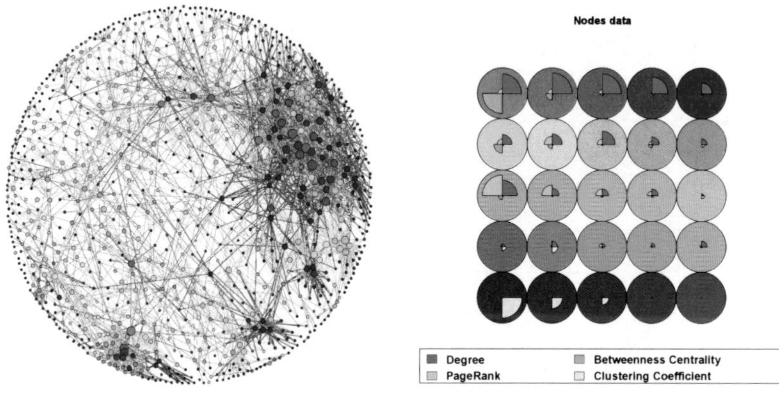

(a) Social Network of Historical Characters (b) Self-organizing Map

Fig. 3 Social Network of Historical Figures. To improve the visualization, we have trained a Self-Organizing Map (SOM), also referred to as Kohonen Map or Kohonen Network [19]. (a) is a visualization where colors correspond to the node's closest classification unit in the Self-Organizing Map (b) and sizes are proportional to the *Degree*. (b) is grid representing a Self-Organizing Map where the nodes are clustered together based on their attributes. Units' background colors matches node colors on the visualization.

are politicians. The top-5 figures with highest PageRanks are Abel Graça, Eduardo Nogueira and Gomes Freire de Andrade. The top-5 by eigenvector centrality is José Antônio Saraiva, Antônio Paulino Limpo, Manuel Alves Branco, Caetano Maria Lopes Gama and Sousa Dantas. A closer look at the dataset and at corresponding Wikipedia articles revealed that some nodes may have be affected by the template used in Wikipedia articles referring to some politicians. At the bottom of those articles there is a list containing the names of all other characters who have occupied one position which that particular person have occupied. It explains the distortions in both PageRank and Eigenvector centralities.

From the ranking of the nodes by Betweenness centrality (Table 2) one can see some interesting information. The first is the fact that the first two are two monarchs, *D. João VI* (John VI of Portugal) and *D. João IV* (i.e. John IV of Portugal). The first was the King of Portugal and Emperor of Brazil who became famous by fleeing to Brazil in 1808 along with all the Portuguese court to escape from the army of Napoleon [20]. The latter was the father of D. Teodósio of Braganza, the first Prince of Brazil. It makes sense that the social network of a monarchical regime has kings and princes among their central characters.

Another historical character who is also noteworthy is Pope Pius VII. In 1804, Pius VII was forced by Napoleon to crown him as the Emperor of France. His presence among the best positioned in the rank of betweenness suggests that the social network was able to identify that Catholic Church was intertwined with many political issues at the time. Also, Pope Pius VII restored the Society of Jesus[6] whose

[6] The Society of Jesus is a religious order of the Roman Catholic Church.

Table 2 Rank of Historical Characters per Betweenness Centrality - The *First Occupation* is the activity a particular person exerted so that he or she became relevant to Brazil. The *Second Occupation* means what activity the person exerted before becoming relevant to Brazil.

Rank	Name	Degree	Role	Occupation
1	D João VI	33	King of Portugal	Emperor of Brazil
2	D João IV	45	King of Portugal	-
3	Pedro de Araújo Lima	39	Politician	Regent of the Empire
4	José Bonifácio de Andrada	38	Statesman	Professor
5	Vasco da Gama	18	Navigator	Explorer

members (the *jesuits*) were responsible for the conversion of the brazilian indians into the catholicism. However, we did not find evidences of direct participation in the history of Brazil. So it is worth further research regarding the actual influence of the Pope Pius VII to the history of Brazil.

5 Conclusions and Future Works

In this paper we have investigated three hypotheses: *(1)* is the network built up from Wikipedia articles a social network? *(2)* who were the most relevant characters for the history of Brazil from the network perspective?, and *(3)* is it possible to use network science to better understand the role that individuals actually played for the course of the history of Brazil?

Although the results shown are preliminary and the work is still in progress, the dataset we have generated has revealed interesting results from both Network Science and History points of view. The network analysis suggests that the network we have built is indeed a social network and displayed both small-world and scale-free characteristics. However, the question of *who was the most important historical character from the network perspective?* remains open. Although the betweenness analysis looks promising more work is necessary to understand the exact importance of betweenness in historical networks. Despite this drawback, we were still able to justify that the betweenness results seem to point to individuals that were central to the development of Brazil.

From the dataset point of view, the next step in this work is to do some community analyses. Preliminary results have shown the emergence of interesting patterns in the communities that will be investigated in depth. Also, it is our intention to start working with Historians since their expertise is fundamental to the validation of results in this context.

Last but not certainly least, we find that the area deserves a special investigation on the network dynamics. From a dynamic analyses, it may be possible to observe the emergence of groups over the time, or to find not only who was the most important historical character now, but how he got to that position; opening a door for prediction of models for future leaders. Such models would help us fulfill the purpose of the study of history which is not only understand the past but also the future.

References

1. Ruffini, G.: Social Networks in Byzantine Egypt. Cambridge University Press (2008)
2. Hook, S.: The Hero in History: A Study in Limitation and Possibility. Transaction Publishers (1945)
3. Borgatti, S.P., Mehra, A., Brass, D.J., Labianca, G.: Network analysis in the social sciences. Science 323(5916), 892–895 (2009)
4. Preiser-Kapeller, J.: Luhmann in Byzantium. A systems theory approach for historical network analysis more. Academia.edu 1316, 1–22 (2012)
5. Preiser-Kapeller, J.: Webs of conversion. An analysis of social networks of converts across Islamic-Christian borders in Anatolia. South-eastern Europe and the Black Sea from the. academia.edu, 1–34 (2012)
6. Wang, C., Kalyanpur, A.: Relation extraction and scoring in DeepQA. IBM Journal of Research and Development 56(3.4), 9:1–9:12 (2012)
7. Berners-Lee, T.: The Semantic Web. Scientific American 21 (2001)
8. Auer, S., Bizer, C., Kobilarov, G., Lehmann, J.: Dbpedia: A nucleus for a web of open data. The Semantic Web (2007)
9. Pattuelli, M.: Linked Jazz: an exploratory pilot. In: Conference on Dublin Core and Metadata Applications 2011, pp. 158–164 (August 2010, 2011)
10. Passant, A., Raimond, Y.: Combining Social Music and Semantic Web for music-related recommender systems. In: Social Data on the Web Workshop (2008)
11. Hienert, D., Luciano, F.: Extraction of historical events from wikipedia, arXiv preprint arXiv:1205.4138 (2012)
12. Cucerzan, S.: Large-scale named entity disambiguation based on Wikipedia data. Proceedings of EMNLP-CoNLL, 708–716 (June 2007)
13. Chang, J., Boyd-Graber, J., Blei, D.: Connections between the lines: augmenting social networks with text. In: Proceedings of the 15th ACM..., pp. 169–177 (2009)
14. Nothman, J., Ringland, N., Radford, W., Murphy, T., Curran, J.R.: Learning multilingual named entity recognition from Wikipedia. Artificial Intelligence 1, 1–25 (2012)
15. Vicentino, C., Dorigo, G.: História Para o Ensino Médio: História Geral e do Brasil, 3rd edn. Editora Scipione, São Paulo (2008) (in portuguese)
16. Watts, D.J., Strogatz, S.H.: Collective dynamics of 'small-world' networks. Nature 393(6684), 440–442 (1998)
17. Chen, Q., Chang, H., Govindan, R., Jamin, S., Shenker, S.J., Willinger, W.: The origin of power laws in Internet topologies revisited. In: Proceedings.Twenty-First Annual Joint Conference of the IEEE Computer and Communications Societies, vol. 2, pp. 608–617. IEEE (2002)
18. Amaral, L.A., Scala, A., Barthelemy, M., Stanley, H.E.: Classes of small-world networks. Proceedings of the National Academy of Sciences of the United States of America 97(21), 11149–111452 (2000)
19. Kohonen, T.: The self-organizing map. Proceedings of the IEEE 78(9), 1464–1480 (1990)
20. Gomes, L.: Como Uma Rainha Louca, Um Príncipe Medroso e Uma Corte Corrupta Enganaram Napoleão e Mudaram e História de Portugal e do Brasil. Editora Planeta do Brasil (1808)

Separate Yet Interdependent Networks: The Structure and Function of European Air Transport

Stephan Lehner

Abstract. Air transport plays a crucial role for connecting people around the globe. Existing network studies often consider only flight networks. However, another important aspect of air transport is the passenger flow from origin to destination – a separate yet interdependent network. This study argues that for a comprehensive study of air transport, both aspects should be taken into account. These networks represent the structure and function of a complex system. The framework of function-structure networks provides a formal model to describe large-scale networked systems. To demonstrate the approach, the global and local properties of the function and structure of European air transport are investigated. This paper also proposes an adjusted betweenness centrality metric and other centrality metrics that can help explain anomalies found in previous investigations of air transportation systems.

1 Introduction

Air transportation plays an important role in modern society. It is the only mode of transportation that can carry people over large distances in an acceptable time. Even for relatively small continents, like Europe, air transport is an enabler of economic prosperity.

Air transport is organized in a decentralized fashion. Many airlines operate various types of networks. Together, these networks constitute regional or global air transport networks. Airlines compete for the favor of passengers on an international market. This indicates that there are two main attributes of air transport. First, airlines operate aircraft between airports. The set of all flights defines the route or flight network. Second, airlines transport passengers from their origin to their destination. This is the passenger flow network.

Stephan Lehner
Institute for Air Transportation Systems German Aerospace Center,
Blohmstr. 18 21079 Hamburg, Germany
e-mail: Stephan.Lehner@dlr.de

G. Ghoshal et al. (Eds.): *Complex Networks IV*, SCI 476, pp. 109–120.
DOI: 10.1007/978-3-642-36844-8_11

In some cases, the flight and passenger flow network are equal. This is called point-to-point transportation. In a hub-and-spoke transportation system, however, aircraft routes and passenger flow differ. Often there is no direct flight between two airports and passengers have to connect at one or more intermediate airports before continuing their journey. Thus, a comprehensive understanding of air transport should take aircraft routes and passenger flow into account.

This study introduces function-structure networks – a multilayer network model for complex systems. The application of function-structure networks enables an investigations of interdependencies between flight and passenger flow networks.

2 Related Work

Air transportation has received a modest amount of interest in the network theory literature. Guimera et al.[5] studied the global flight network. They found that the nodes with the most connections are not necessarily the nodes with the highest betweenness centrality. Barthelemy et al.[1] highlighted the need to include link strength in considerations of complex systems in general and provided examples for air transport. DeLaurentis[4] and Bouonova[2] study global and local network properties of the US airline networks, and Li and Cai[8] investigate the topological properties of the airport network in China. Wuellner et al.[13] discuss the resilience of US passenger airline networks.

Europe is another important air travel market, yet it has drawn much less attention in the network theory literature. Some economic studies[10][3] discuss centralization aspects of air transport in Europe.

Previous investigations of air transport have considered only flight networks, but the *raison d'être* of an commercial air transportation system is to transport passengers from their origin (the airport where a passenger starts its aerial journey) to the final destination (airport where a passenger terminates its aerial journey). The existence of indirect travel paths make airline flight networks and passenger flow networks separate yet interdependent networks. It thus appears necessary to account for multiple linkage types.

Multiplex networks are well known in the social network literature. They describe different forms of ties between actors[12]. A recent investigation[7] from the physics literature proposed the use of networks on multiple layers to understand a system. These approaches consider the interdependence of different attributes of a single system. Thus, multiplex networks are well suited to compare flight networks and passenger flow of air transport.

3 Function-Structure Networks

To illustrate the idea of interdependent system networks, consider the illustrative flight and passenger networks in Figure 1 as an example.

Both flight networks are equal. However, they can provide different passenger flows. In the left networks, passengers travel only from the outer airports to the

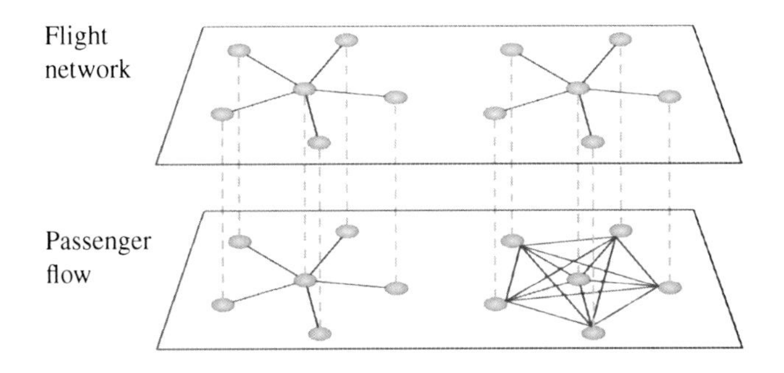

Fig. 1 Illustrative example of the interdependence of structure and function networks

central airport. This could be the case when the central airport is close to a city that is a major cultural or economic center. In the right networks, in contrast, passengers travel between all airports by connecting at the central node. In this case, the airline has to coordinate its flight plan so that connection time for passengers is within acceptable boundaries.

From a systems point of view, the flight network and the passenger flow network represent two main attributes of a system. Enabling passengers to travel from their origin to their destination is the purpose of the system. Thus, passenger flow represents a *function* of the air transport system. There are several possibilities of how this function could be provided. The system *structure or form* characterizes the means by which the function is provided.

The main point of Figure 1 is that even if the structures of two systems are equal, the function could be significantly different. Therefore, it appears to be desirable for certain systems to understand structure and function as a coherent whole. Rather than asserting that function follows form, or vice versa, this study understand function and structure as mutually interdependent.

Note that the notion of function used in this study differs somewhat from other studies[9][6] where function is often understood as a dynamic network process.

A function network G^F is a model of the system that describes its purpose by means of a network. A function link L^F with weight or strength l_{ij}^F connects two nodes in a function network. It thus indicates an atomic functional relationship between two system elements. The network composed of function links represents the collectivity of all atomic functions a system can provide. For air transport, this network indicates how many people start their travel at a node i and end their journey at a node j.

A structure network G^S can model *how* a system provides its functions by accordingly connecting the system elements with structure links L_S , each having a weight l_{ij}^S. A path in this domain is called a structure path. In air transport, the structure network consists of all considered flights.

Measures that are valid for both networks include the number of links $|L^F|$ and $|L^S|$ and the cumulative link strength (sum of all link strengths) $|W^F|$ and $|W^S|$. Since the adjacency matrix of both networks is almost symmetric, this study uses only the in-degrees as denoted by q_i^F and q_i^S for a node i, and $\langle q^F \rangle$ and $\langle q^S \rangle$ for the mean node degree.

A function- structure (F-S) network is a model to mutually describe the function and structure of a system. It is the union of a system's function and structure network. The set of nodes in both networks are equal, i.e. $N = N^F = N^S$. Since there are (at least) two different link types, it is a multi-relational or multiplex network. The number of links $|L^F|$ and $|L^S|$ may (and typically does) differ. The same is true for the cumulative strengths for the function and structure network, respectively.

4 Application of Function-Structure Networks: European Air Transport

4.1 Data Source

The results presented in this paper are based on the Sabre Airport Data Intelligence database[11]. This global market database is widely used by airports, airlines and aircraft manufacturers. Typically, this data changes frequently to reflect the most recent updates. However, the data used here was declared as *final* by the database provider.

The database collects the number of passengers for a given origin- destination market from online booking systems and other sources. The function network is constructed from *demand passengers* data. The number of passengers that actually flew between two airports is known as *segment passengers* data and is the source for the structure network.

From this data, the segment (i.e. structure) and demand (i.e. function) networks of the 12 largest European airlines are extracted. Additionally, this study investigates the intra-European air travel network (i.e. start *and* end node of each link are in Europe). This network is not the union of all European airline networks, but rather includes only those subnetworks that are within Europe. All data entries for flights and origin-destination markets in the year 2011 are considered. For consistency, we only include links that have a link strength of 600 passengers (50 passengers per month) and greater.

4.2 Fundamental Network Properties

Table 1 shows the main properties of the structure and function networks of various European air transport networks.

The number of nodes $|N|$ ranges from 86 for Alitalia to 199 for Turkish Airways. British Airways (BA) and Ryanair have the same number of nodes. However, the number of structure links $\left|L^S\right|$ reveals that Ryanair offers six times as many flight links as BA. The number of function links in the Ryanair network is about the same as the structure links. Despite having fewer flight links, the BA network, provides much more function links. This is an indications of the network strategy

Table 1 Network Properties of different European air transport networks. The name indicates the IATA abbreviations of the considered airlines. These are: Air Berlin (AB), Air France (AF), Alitalia (AZ), British Airways (BA), Ryanair (FR), Iberia (IB), KLM (KL), Lufthansa (LH), Swiss (LX), SAS (SK), Turkish Airlines (TK), easyJet (U2) and the intra-European air travel network (Europe).

Name	$\|N\|$	$\|L^S\|$	$\|L^F\|$	$\frac{\|L^S\|}{\|L^F\|}$	$\frac{\|W^S\|}{\|W^F\|}$	ρ^S	ρ^F	$\frac{\rho^S}{\rho^F}$	G^S	G^F	$\frac{G^S}{G^F}$
AB	155	1110	1671	0.664	1.094	0.034	0.058	0.588	0.833	0.833	1.001
AF	171	493	4040	0.122	1.450	0.005	0.139	0.038	0.768	0.740	1.037
AZ	86	276	1913	0.144	1.406	0.015	0.263	0.056	0.780	0.741	1.054
FR	177	2668	2680	0.996	1.000	0.075	0.076	0.995	0.690	0.690	1.000
BA	177	438	3531	0.124	1.426	0.003	0.107	0.026	0.760	0.721	1.055
IB	97	191	2316	0.082	1.549	0.000	0.243	0.000	0.771	0.698	1.105
KL	131	258	3514	0.073	1.847	0.000	0.207	0.000	0.707	0.611	1.157
LH	191	667	7304	0.091	1.646	0.008	0.204	0.039	0.795	0.737	1.078
LX	102	264	2162	0.122	1.491	0.006	0.194	0.032	0.764	0.704	1.084
SK	93	349	1633	0.214	1.376	0.020	0.173	0.114	0.791	0.761	1.040
TK	199	540	3873	0.139	1.622	0.004	0.091	0.041	0.748	0.698	1.072
U2	127	1051	1057	0.994	1.001	0.051	0.051	0.993	0.713	0.713	1.000
Europe	711	15477	25405	0.609	1.189	0.028	0.049	0.569	0.824	0.809	1.019

of both airlines. British Airways routes many of its international passengers through its London Heathrow hub. Ryanair, in contrast, offers point-to-point transportation and provides almost exclusively direct connections.

Thus, there is an interesting relation between function and structure on the one hand and *directness* on the other. In fact, it could be stated that quantifying *directness* can also quantify the interdependence of function and structure of complex systems.

Dividing the number of structure links by the number of function links, i.e.

$$D_1 = \frac{|L^S|}{|L^F|} \tag{1}$$

is probably the simplest directness measure. It corresponds to the topological directness because it uses only topological information from the structure and function network. In cost-constrained networks (like air transport), there is an incentive to keep the number of structure links as low as possible. Thus, in the most expensive case, there is a structure link for each function link and $D_1 = 1$ (see Ryanair and Easyjet). For values significantly smaller than 1 (say below 0.25), the networks have a high indirectness. In these F-S networks, there are only few structure links and many function links. Air Berlin and Intra-European air travel represent mixed direct-indirect networks.

Saving on flight links, however, comes with a price. Passengers have to be transported somehow, and thus the cumulative link strength has to increase. A passenger who has to change flights during his travel requires at least two seats (one on the first segment and one on the second and so on). Thus, in an indirect F-S network, there is an additional operating expenditure for the airline. Dividing the cumulative structure link weight by the cumulative function link weight, i.e.

$$D_2 = \frac{|W^S|}{|W^F|} \tag{2}$$

quantifies this increased effort. For air transport, this measure (i) represents the average number of hops a passenger has to fly and (ii) is a relative measure of how many excess seats an airline has to provide. The lowest expense (valid for Ryanair and Easyjet) is $D_2 = 1$ if the network has exclusively direct connections. Air Berlin and intra-European air travel are mixed networks. Airlines with a low D_1 value have an excessive number of seats.

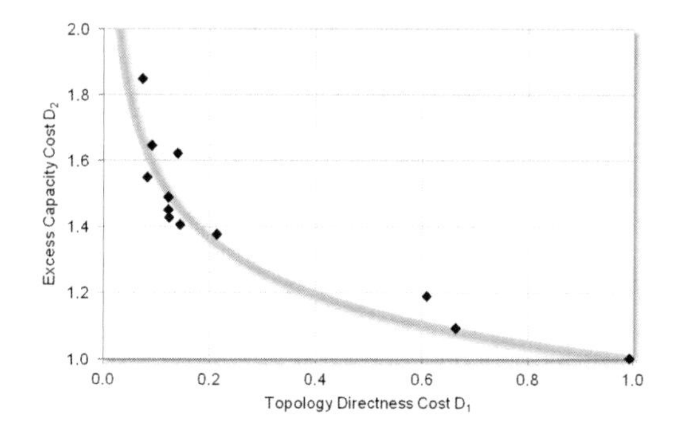

Fig. 2 Topology Directness Cost versus Excess Capacity Cost. The best fit through the data was provided by a power law and suggests that a trade-off has to be made.

Both, D_1 and D_2 are indicators of different cost types for operating either indirect or direct networks. Table 1 shows that if an airline has low cost in D_1, it has high cost in D_2 and vice versa. Figure 2 visualizes this relation. A power law provided the best fit for the data. There appears to be a trade-off between these two cost types. The course of the fit could suggest that there exists some kind of Pareto efficiency relationship. There are relatively few airlines that operate in the range $D_1 = 0.2 \ldots 0.8$, yet this region appears to provide a significant reduction in topology directness cost with only a modest increase in excess capacity cost. Instead, most airlines either opt for a high D_1/low D_2 or high D_2/low D_1. However, the number of data points is limited here. Including more airlines (e.g. from other continents) could make the statement on this trend more robust.

The density ρ^S of the structure networks reveals that none of the airline flight networks is dense. This comes as a surprise because flight networks of point-to-point transportation providers, like Ryanair and Easyjet, are often attributed as being almost fully connected. However, it should be acknowledged that directness as an indicator of point-to-point characteristics depends on the interplay of two networks (structure and function) whereas density is a property of the structure network alone.

Considering Fig. 1, it is possible that a star network provides only direct travel. After all, point-to-point, in the purest sense of the words, just indicates direct travel and does not make any further assertions. Compared to more indirect networks, however, the point-to-point philosophy of some European airlines generates significantly (factor 2-10) denser networks. As expected, the density of function networks is more pronounced in indirect networks. Also not surprisingly, the relation of the two densities $\frac{\rho^S}{\rho^F}$ shows a similar behavior as the topological directness D_1.

Another interesting network property is how centralized the structure network is. In the transportation literature, this centralization has been measured by the Gini coefficient, a metric that indicates the inequality of a frequency distribution. For weighted networks, it is defined as in Ref. [3]

$$G = \frac{\sum_i \sum_j |q_i^S - q_j^S|}{2\,|N^S|^2 \langle q^s \rangle} \tag{3}$$

A Gini coefficient of 1 would correspond to a maximally centralized network. Table 1 shows that, according to this index, the centralization of all networks is similar. In particular, the direct Easyjet network appears to be somewhat more centralized than the indirect KLM network. Again, this is in contrast to many assertions that claim that point-to-point transportation, in general, results in flight networks that are not centralized. In networks that provide point-to-point transportation, there can exist few central nodes that have a high attractiveness. For example, in the Ryanair network, Dublin and London-Stansted have a high degree and are also attractive business and leisure destinations and many passengers terminate their travel at one of these airports. The Gini coefficient of the function networks is only slightly smaller than the same index for the structure network. Thus, the the Gini coefficients alone provides only limited insight into the interdependence of function and structure of complex systems.

4.3 Degree Distributions

After having investigated the fundamental network properties of the structure and function of European air transport, the next step is to consider a more in-depth analysis of how the nodes are connected. Figure 3 plots the cumulative in-degree and incoming strength distribution functions for the function and structure networks of four European airlines and intra-European air travel. The four airline networks in Fig. 3 are representative of all 12 European airlines that Table 1 contains.

Air Berlin (Fig. 3a)) has already been identified as the sole mixed direct-indirect airline network. The best fit for the cumulative degree distributions of its flight and passenger flow network is a power law. Similarly, the cumulative strength distribution is also close to a power law, although not as pronounced as the degree distribution. There is a small difference between the distribution functions of the structure and function network, indicating that there is some indirectness.

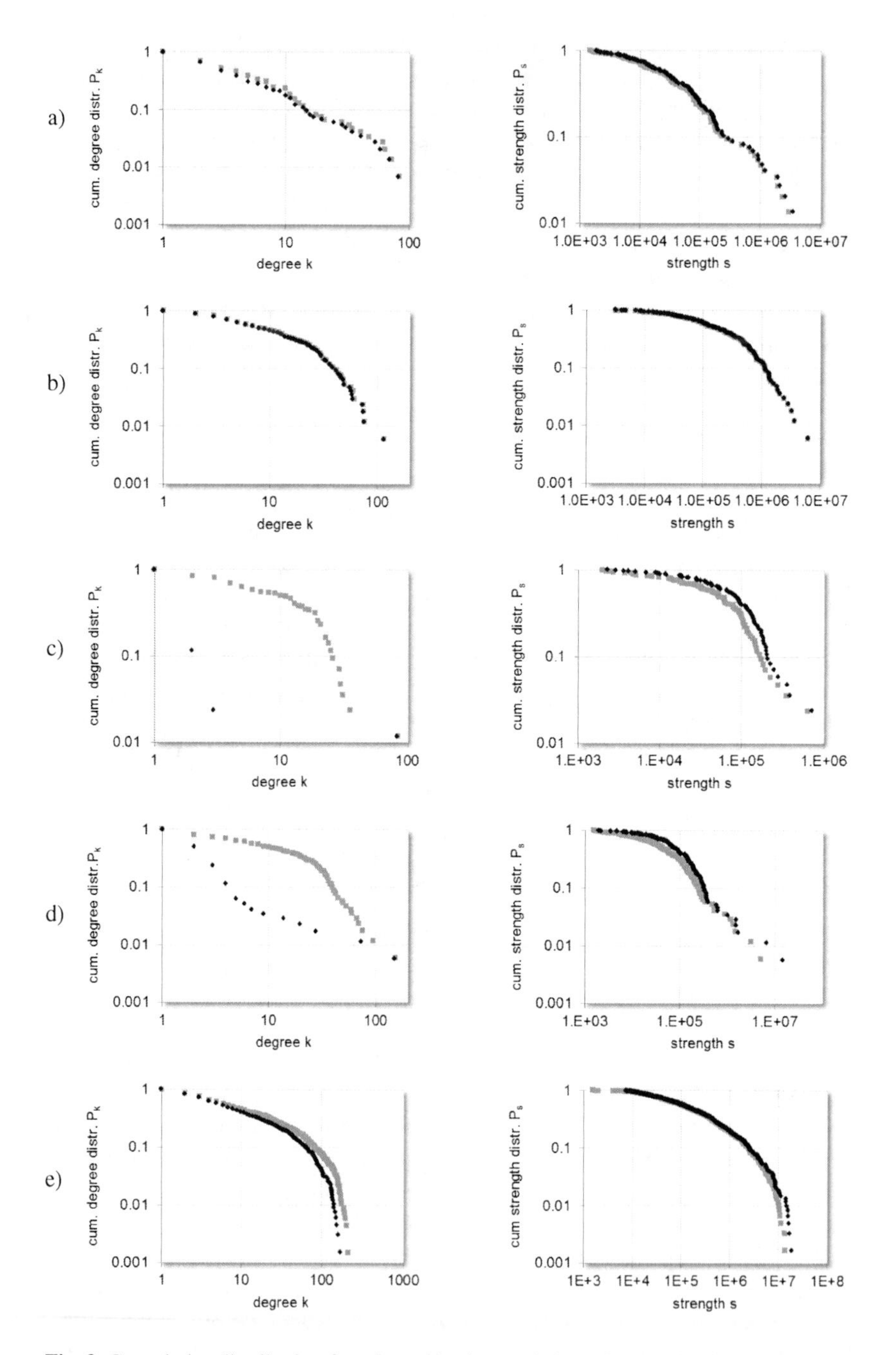

Fig. 3 Cumulative distribution function of in-degree (left) and strength (right). The black dots indicate the structure (i.e flight network) and the gray dots represent the function (i.e. passenger flow network). The networks represent a) Air Berlin, b) Ryanair, c) Iberia, d) Lufthansa and e) intra-European air travel.

The cumulative distributions of Ryanair in Fig. 3b) are similar to those of Easyjet. This type of air transport network shows no difference between function and structure, highlighting its high directness. The best fit for the distributions is a power law with an exponential decay. This is another indication that this network is neither fully connected or nor decentral.

The network characteristics of Iberia (Fig. 3c)) are similar to those of KLM and British Airways. In the structure network, there is one central node where most flight links originate or terminate. The degree distribution of the passenger flows can be approximated by a two-regime power law (one for nodes with few links and one for those with many links). The cumulative strength distributions of structure and function can also be fitted by a truncated power law.

Lufthansa's network distributions in Fig. 3d) are similar to those of Air France, Alitalia, Swiss, SAS and Turkish Airlines. These airlines operate indirect networks but the distribution is somewhat different to those of Iberia in Fig. 3c). Here, the cumulative degree distribution of the structure network follows roughly a two-regime power law. The cumulative strength distributions show a truncated power law behavior.

The cumulative degree distributions of the intra-European air travel networks in Fig. 3e) show also a truncated power law behavior. This is in line with the observations of Guimera et al.[5] for the world wide air transport network. The cumulative strength distributions can also be fitted by a power law, although the goodness of fit is not as high as for the the degree distribution.

4.4 Nodal Centrality

As indicated before, previous studies have found an anomalous behavior in air transport networks. Guimera et al.[5] show that in the global flight network, nodes with the most connections are not necessarily the nodes with the highest betweenness centrality. It appears interesting to investigate nodal centrality when both, the flight network and passenger flow network are taken into account.

The betweenness centrality (BC) of node i is defined as the number of shortest path between all nodes where i is a part of. BC assumes a potential relationship between each node. It thus indicates how important a node is to facilitate total connectedness. BC can be understood as a fully connected network that acts as a driving force between all nodes and is comparable to a function network. As illustrated, some function networks might not be fully connected but rather show complex behavior. If this is the case, the calculation of betweenness centrality should take this configuration into account. Thus, this study proposes the use of a function adjusted betweenness centrality (FAB) measure. It counts only the number of times a node i is part of a shortest structure path between node s and node t which are connected by a function link. It is defined as

$$b_i = \sum_{st} n_{st}^i \quad \forall\; l_{st}^F \tag{4}$$

A similar measure has also been proposed by Kurant[7]. Previous studies have normalized betweenness centrality by the average betweenness centrality $\overline{b}$. Figure 4 plots both, normalized betweenness centrality and normalized function adjusted betweenness centrality over nodal degree for all airports in the intra-European travel network.

The betweenness centrality plot shows nodes that have a low nodal degree and high betweenness centrality. A particular node with low degree/ high centrality is Tromso airport, which is a relatively small airport in northern Norway. Its high betweenness centrality results from its connections to Russian airports. However, it can at least be questioned that Tromso is actually used as a connecting airport.

The plot for function adjusted betweenness centrality provides a less scattered correlation. For this metric, there is no node that has a low degree and high betweenness centrality. This indicates that taking system structure and function into account can explain some of the anomalous behavior found in previous studies. There are nodes that would have a high betweenness centrality if the function network would be fully connected. However, as the function network typically is not fully connected, adjusting betweenness centrality can help in explaining real-world peculiarities. Nevertheless, one anomaly remains: there are nodes that have a high degree and low betweenness centrality.

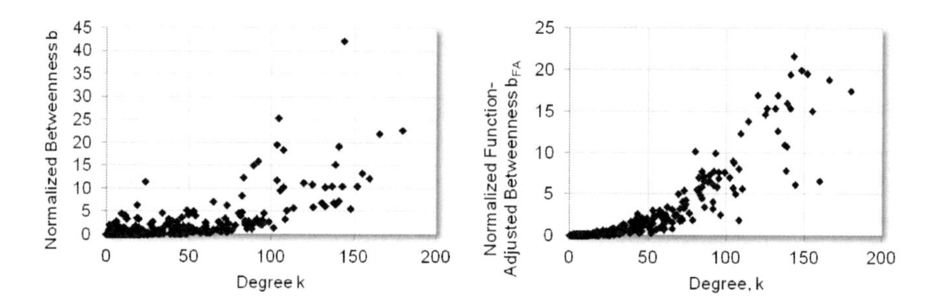

Fig. 4 Normalized Betweenness and normalized function adjusted betweenness versus nodal degree for the intra-European travel network

Therefore, it appears appropriate to distinguish between structure hubs, function hubs and transfer hubs. A *structure hub* node has many incoming structure links q_i^S or a high incoming structure strength $c_i^S = \sum l_{ij}^S$. In air transport, this means that many passengers arrive at an airport, but not necessarily terminate their journey. Some of them might connect to another flight and continue their journey.

A function hub, on the other hand, has many incoming function links q_i^F or a high incoming function strength c_i^F. This nodes represent the most important destinations for passengers.

A third centrality metric for function-structure networks is *transferness*. This metric indicates to what extent passengers connect at an airport. Function adjusted betweenness centrality is a metric that indicates a *transfer hub* in an unweighted network. For weighted networks, the transferness score is simply the difference between the incoming structure strength and function strength

$$c_i^T = \sum l_{ij}^S - \sum l_{ij}^F \tag{5}$$

at a given node i. Table 2 highlights the top ten structure, function and transfer hubs for in-degree and incoming link strength. The degree and strength score are normalized by total degree and total strength, respectively.

Table 2 Structure hubs, function hubs and transfer hubs of the intra-European air travel network. The name indicates the airport's IATA code

Structure Hubs		Function Hubs		Transfer hubs	
Degree	Strength	Degree	Strength	FAB	Transferness
PMI 0.013	MAD 0.031	FRA 0.010	BCN 0.027	FRA 0.029	FRA 0.089
BCN 0.011	FRA 0.028	CDG 0.010	MAD 0.026	PMI 0.029	AMS 0.060
STN 0.010	LHR 0.027	AMS 0.010	LHR 0.022	BCN 0.028	MUC 0.059
LGW 0.010	CDG 0.026	BCN 0.010	LGW 0.022	MUC 0.028	CDG 0.057
AMS 0.010	BCN 0.026	MUC 0.009	CDG 0.020	CDG 0.026	MAD 0.057
MUC 0.010	AMS 0.025	LHR 0.009	FCO 0.020	AMS 0.026	LHR 0.050
FRA 0.010	FCO 0.024	FCO 0.009	PMI 0.019	MAD 0.023	FCO 0.043
TFS 0.010	MUC 0.024	DUS 0.009	AMS 0.018	CPH 0.022	ZRH 0.035
DME 0.010	LGW 0.019	BRU 0.009	STN 0.017	TFS 0.022	VIE 0.030
MAN 0.010	OSL 0.018	PMI 0.009	MUC 0.017	FCO 0.022	CPH 0.029

Several observations are noteworthy. For example, Frankfurt (FRA) is the second most central structure hub in terms of link strength, yet in terms of number of passenger that terminate their journey, it does not appear in the top 10. Frankfurt has the highest number of connecting passengers for inner-European travel and is thus important for indirect travel.

Table 2 can also help explain the phenomenon of high degree/ low betweenness centrality nodes in Figure 4. Consider for example London Stansted (STN). In Fig. 4, it appears in the lower right region of the FAB plot. This airport is a structure degree hub and function strength hub. However, there are only few connecting passengers at these airports. This comes at no surprise as Stansted is the home base of Ryanair, an airline that operates a direct F-S network. Thus, the introduction of the three hub categories has helped in explaining the role of nodes in a given network.

5 Conclusions

This study has introduced function-structure networks. Several European air transport networks were investigated, where passenger flow represents function and the flight network characterizes system structure.

Two measures that describe function-structure interdependence resemble the notion of directness. Real world data suggests that these measures are conflicting, a finding that could motivate further research on how systems evolve.

The results indicate that function-structure directness and nodal centrality should be understood as two separate notions. Some air transport networks showed a power-law behavior even if they are often characterized as point-to-point networks. Using passenger flow as the system's function, the betweenness centrality measure was adjusted to better represent the real world behavior of air transport. Thus, including system function considerations can help in better explaining system peculiarities.

It would be interesting to compare the properties of the intra-European air travel network and the global air transport system. Network formation models and resilience studies of large scale network systems could also benefit from including considerations on on function- structure interdependency.

References

1. Barrat, A., Barthélemy, M., Pastor-Satorras, R., Vespignani, A.: The architecture of complex weighted networks. PNAS 101, 3747–3752 (2004)
2. Bounova, G.A.: Topological Evolution of Networks: Case Studies in the US Airlines and Language Wikipedias. Dissertation, MIT (2009)
3. Burghouwt, G., Hakfoort, J., Ritsema van Eck, J.: The spatial configuration of airline networks in Europe. Journal of Air Transport Management 9, 309–323 (2003)
4. DeLaurentis, D., Han, E.-P., Kotegawa, T.: Network-Theoretic Approach for Analyzing Connectivity in Air Transportation Networks. Journal of Aircraft 45 (2008)
5. Guimera, R., Mossa, S., Turtschi, A., Amaral, L.: The Worldwide Air Transportation Network: Anomalous Centrality, Community Structure, and Cities Global Roles. PNAS 102, 7794–7799 (2005)
6. Cohen, R., Havlin, S.: Complex Networks: Structure, Robustness and Function. Cambridge University Press (2010)
7. Kurant, M., Thiran, P.: Layered Complex Networks. Physical Review Letters 96 (2006)
8. Li, W., Cai, X.: Statistical analysis of airport network of china. Phys. Rev. E 69 (2004)
9. Newman, M.E.J.: The Structure and Function of Complex Networks. SIAM Review 45 (2003)
10. Reynolds-Feighan, A.: Traffic distribution in low-cost and full-service carrier networks in the US air transportation market. Journal of Air Transport Management 7, 265–275 (2001)
11. Sabre. Airport Data Intelligence Database (2012), `http://www.sabreairlinesolutions.com/home/`
12. Wassermann, S., Faust, K.: Social Network Analysis: Methods and Applications (Structural Analysis in the Social Sciences. Cambridge University Press (1994)
13. Wuellner, D., Roy, S., D'Souza, R.: Resilience and rewiring of the passenger airline networks in the United States. Physical Review E 82 (2010)

Evaluating the Stability of Communities Found by Clustering Algorithms

Tzu-Yi Chen and Evan Fields

Abstract. Since clustering algorithms identify communities even in networks that are not believed to exhibit clustering behavior, we are interested in evaluating the significance of the communities returned by these algorithms. As a proxy for significance, prior work has investigated stability: by how much do the clusters change when the network is perturbed? We describe a cheap and simple method for evaluating stability and test it on a variety of real-world and synthetic graphs. Rather than evaluating our method on a single clustering algorithm, we give results using three popular clustering algorithms and measuring the distance between computed clusterings via three different metrics. We consider the results in the context of known strengths and weaknesses of the three clustering algorithms and also compare our method against that of measuring modularity. Finally we give evidence that our method provides more information than does the modularity.

1 Introduction

Clustering algorithms are used in a range of fields to identify meaningful groupings of entities. The input is a graph whose nodes and edges represent entities and the connections between them respectively, and the output is an mapping of nodes to groups such that nodes in the same group are more connected to each other than they are to nodes in other groups. For example, in the network studied in [27], nodes represent court cases and an edge between two nodes means that one case cited the other. Since cases on similar topics are more likely to cite each other, the clusters identified in that network might be expected to reflect different areas of law.

However, clustering algorithms are only useful for understanding real-world networks if the groupings that they return are significant. In other words, the clusters detected should reflect true underlying group structure in the network rather than

Tzu-Yi Chen · Evan Fields
Computer Science Department, Pomona College, 185 E. Sixth St, Claremont, CA 91711
e-mail: tzuyi@cs.pomona.edu, ejf12009@mymail.pomona.edu

G. Ghoshal et al. (Eds.): *Complex Networks IV*, SCI 476, pp. 121–132.
DOI: 10.1007/978-3-642-36844-8_12

effects due to noisy or incomplete data. Unfortunately, significance is difficult to evaluate. One approach is to use a proxy measure such as the *modularity* of a clustering, defined in [31], though we note all such measures will be imperfect [22]. In the case of modularity, which remains very popular, it is known to have trouble with certain random graphs [17] as well as some real-world networks [16].

An alternative to using a proxy measure is to evaluate significance by looking at the *stability* or *robustness* of a clustering: by how much do the clusters change if the input graph is slightly perturbed? A standard approach to answering this question is to compute the clustering of the initial network, compute clusterings of perturbed versions of the network, and then evaluate the difference between the new and original clusterings. The belief is that if the initial output does not consist of significant clusters, perhaps because the graph itself does not have significant clusters, then small perturbations should be more likely to lead to very different clusterings. Several previous studies, discussed in Section 2, have done exactly this. They choose a method for perturbing the initial network, a clustering algorithm, and a measure of distance. They then modify a parameter which controls the magnitude of the perturbation and analyze plots of the distance as a function of the size of the perturbation.

In this paper we propose a simpler perturbation strategy which only adds a single edge to the network and thus removes the parameter controlling the magnitude of the perturbation. Even though this method is less computationally expensive than previously proposed techniques, we demonstrate that it still reveals interesting differences in clustering behavior.

Furthermore, previous studies are limited by the fact that they propose a method for perturbing, clustering, and plotting, but their experimental results typically use only a single clustering algorithm and a single measure of distance. They then typically compare the plots on random graphs to plots on networks with community structure. By way of contrast, as described in more detail in Section 3, we report on experimental results from using our method with three different popular clustering algorithms and three different measures of cluster similarity.

The hope is that these experiments reveal qualitative or quantitative differences between the networks in our test suite that are independent of clustering algorithm and distance metric. Ultimately the goal is to determine not just how stable a particular clustering is, but whether a network exhibits underlying clustering structure. This study can be viewed as a small step towards separating the properties of a network from the properties of the algorithms used to understand it.

Our test suite consists of real-world networks that are believed to exhibit clustering behavior, meshes which are generally felt not to exhibit as much clustering behavior, and random graphs. We evaluate simple statistics of the distances between clusters and find, for example, that the stability of the clusterings computed on an Erdos-Renyi random graph are markedly different from that of clusterings computed on real-world networks. Indeed, this is true for all three clustering algorithms and all three distance measures tried. We also find examples where known weaknesses in clustering algorithms reflect themselves in the stability of the clusters, as measured using our protocol. Furthermore we find examples where the stability of a clustering is not predicted by its modularity.

2 Related Work

As noted previously, the standard approach to evaluating the significance of graph clusters involves repeatedly perturbing the network then evaluating the distances between the clusterings of the original and perturbed networks. We summarize some of this work here and note differences from our approach, but also emphasize that one of the contributions of our work is simply the comparison of results using different clustering algorithms and measures of clustering similarity.

In [15] Gfeller et al propose randomly perturbing edge weights and seeing how the cluster structure changes as a result. Their method repeatedly perturbs the edge weights, then looks at the fraction of times that each pair of nodes are classified in the same cluster and combines this into a single global measure of sensitivity that they call "clustering entropy". The results in their paper use the MCL clustering algorithm [10]. In [33] Rosvall and Bergstrom also propose perturbing the edge weights and then evaluating the resulting clusters, though their focus is on visualizing change over time. The results in [33] use the clustering algorithm described in [32]. The work in our paper, on the other hand, looks at perturbations to the structure of the network rather than to the edge weights.

In [21] Karrer et al propose perturbing the initial network by moving each edge, with probability p, to a new location between vertices i and j with probability proportional to the product of the degrees of i and j in the original graph. The perturbed networks maintain not only the number of edges and vertices, but also the expected degree list. The method in [21] then plots the distance between the clusterings of the original and perturbed networks as a function of p. They evaluate their method using the spectral clustering algorithm in [30] and the variation of information similarity measure [26]. Their results are inconclusive: they find that the plots for some networks which are believed to have community structure do not appear significantly different from the plots for random networks. While their procedure requires varying the paramater p and generating multiple networks for each value of p, our method is less computationally expensive because it only ever adds a single edge. An additional advantage is that adding edges never disconnects a connected network.

In [19] Hu et al extend the work of Karrer et al. While their perturbation method also moves a percentage of the edges p, it does not maintain the same expected degree list. More importantly, while they also plot the distance between the clusters as a function of p, Hu et al propose a measure of plot similarity R which eliminates the effects of scale. If G is the original network and $G(p)$ is the perturbed network, they define

$$R = \int_0^1 E[I(G,G(p)) - I(G_r,G_r(p))]dp \qquad (1)$$

where $I(G,G(p))$ is the similarity between the original and perturbed clusterings and $I(G_r,G_r(p))$ is the similarity between randomly chosen equivalent clusterings in equivalent random graphs. By "equivalent" they mean that G_r has the same number of edges, the same number of vertices, and the same size clusters as G (and similarly for $G_r(p)$ and G_r). The experiments use the same spectral clustering algorithm used in [21], but use a distance measure based on the normalized mutual information (NMI) index [8]. The authors observe that the results are not affected by the simpler edge perturbation method and that the R measure suggests that social networks have more significant communities than networks from various biological sources. In addition, they find that the R measure is "roughly independent of the network size and group number." While our method does not directly address scale, it has the advantage of using both a simpler perturbation strategy and an analysis based on simple statistics. Furthermore, the measure R does not distinguish between differences at large perturbations and those at small perturbations, which may obscure facts about the original network.

In [27] Mirshahvalad et al also perturb an initial network by adding edges, which they note is good for sparse networks since it keeps from fragmenting the network. For the results in their paper, edges are added based on the triangle completion model, clusters are computed using the algorithm in [32], and distances are compared using the NMI. They test their methodology on random graphs, but the focus is on an in-depth study of a network created from citations between court cases.

While not the focus of this paper, there has also been related work in evaluating the stability of distance-based clusterings. Here data points are mapped to a location in space and then grouped into clusters such that objects in the same cluster are closer to each other than to those in other clusters. For example, in [2] and [24] researchers propose evaluating the stability of clusterings of biological data by looking at the distance between clusterings of subsamples of the entire dataset.

Finally we note that several earlier studies have compared clustering methods, although none has done so in the context of the stability of their outputs [23, 28, 34].

3 Methodology

Given a network, we first compute its clusters. We then add a randomly chosen edge sampled using preferential attachment and find the clusters in the new network using the same clustering algorithm. Finally we compute both the modularity of the clustering and the distance between the clusterings on the original and perturbed networks. We repeat the sampling, clustering, and computation of distance and modularity 100 times. The same edge is never resampled, so all 100 perturbed networks are distinct.

3.1 Clustering Algorithms

Of the many clustering algorithms available (see [14]), we chose the following three.

- the Clauset-Newman-Moore (CNM) algorithm [6], which is an optimized version of the greedy agglomerative hierarchical clustering technique described in [29]. Each vertex starts off in its own cluster and, at each step, the two clusters that are most connected are merged into a single cluster. The CNM algorithm computes the modularity after each merge and outputs the clustering at the point where the modularity is maximized. We used the implementation at [5].

- the Louvain clustering algorithm [3], which assigns each node to its own cluster and then moves nodes to the neighboring cluster which most increases modularity until a local maximum is obtained. Each cluster is then collapsed into a single node and the process repeated until no further improvements to the modularity can be found. The code is available at [4].
- the Markov Cluster Algorithm (MCL) [10], which is based on simulating flow diffusion in a graph. The algorithm repeatedly squares the transition matrix, then adjusts the matrix to emphasize differences between vertices and to keep computational costs low. The magnitude of the adjustment is determined by an inflation parameter which we fixed at 1.4. The code is available at [12].

3.2 Distance Measures

In [26] measures of clustering similarity are described as being based on pair counting, cluster matching, or information theoretic measures. We used one of each type:

- the Jaccard distance, which is a pairwise comparison of vertices to see how many are grouped together in both clusterings [20].
- the Split-Join [11] distance, which measures the number of vertices that must be reclassified to turn one clustering into the other. For these experiments the split-join distance was computed using the igraph library [7]; the result was normalized to be between 0 and 1.
- the normalized mutual information (NMI) [8] similarity measure is an information theoretic measure which considers the information content of the clusters. The NMI was calculated using a function in the igraph library. We report on 1 minus the NMI measure, since NMI measures similarity rather than distance.

3.3 Test Suite

Table 1 summarizes the networks used for the results in this paper. The first four are real-world networks that are taken from a variety of sources but which are all generally believed to exhibit clustering behavior; these networks are all included in the DIMACS10 challenge set (`http://www.cc.gatech.edu/dimacs10`). The next four are synthetic meshes: of those the first three are also included in the DIMACS10 set, the last is a regular mesh and is available from [9]. The last three are generated using three common random graph models: the Erdos-Renyi model, the Barabasi-Albert model, and the Watts-Strogatz model. As we discuss in the following paragraph, each of these models is expected to generate graphs with different degrees of clustering. The ba-10000 and ws-10000 matrices were generated using the networkX package [18].

The Erdos-Renyi model, which starts with n vertices and then adds the edge (i, j) with probability p [13], makes the assumptions that the probability of each edge

Table 1 Networks in the test suite

name	nodes	edges	description
de2010	24115	116056	adjacent census blocks in Delaware
pgp	10680	48632	largest connected component of pgp users
as-22july06	22963	96872	structure of internet routers
power	4941	13188	topology of the Western States Power Grid
del-n10	1024	6112	delaunay triangulation of random points in the plane
del-n11	2048	12254	delaunay triangulation of random points in the plane
del-n12	4096	24528	delaunay triangulation of random points in the plane
gr30_30	900	7744	30x30 discretized grid
er-5000	5000	50000	graph generated by Erdos-Renyi model, average degree 10
ba-10000	10000	99900	graph generated by Barabasi-Albert model, average degree 10
ws-10000	10000	50000	graph generated by Watts-Strogatz model [35], average degree 5, 30% probability of moving an edge

existing is equally likely and that edges are independent. Such assumptions lead to the Erdos-Renyi model generating graphs with degree distributions that converge to a Poisson distribution and that have little clustering. But real-world networks tend to have different behavior. First, graphs such as the World Wide Web network and citation networks are known to be scale-free, which means their degree distribution follows a power law. The Barabasi-Albert model generates scale-free graphs by starting with a small connected graph and then iteratively adding nodes. As each node is added, edges are added between it and the existing nodes using a preferential attachment model: nodes with higher degree are more likely to be connected to the newly added node [1]. Second, real-world networks often do exhibit (local) clustering.

The Watts-Strogatz model creates networks with clustering behavior as well as short average path lengths by starting with a regular ring lattice and then moving each edge with some probability β.

While data has been gathered on a larger set of networks, including multiple instances of networks generated using the three random graph models, the results in this paper use only the 11 in Table 1. Also note that all 11 networks consist of a single connected component; in some cases a random graph generator was run multiple times to get such a network.

4 Results

In this section we first analyze simple statistics on the distribution of distances between the original network and the 100 perturbed networks for each combination of clustering algorithm and distance measure. We then examine the modularity values of the clusterings computed by each algorithm.

As noted earlier, the expectation is that if the clusters found are not significant, then the modularity should be small and the distances between the clusters of the perturbed network and the original network should be greater. We generally expect the meshes to exhibit less significant clustering than the graphs from real-world networks. We would expect the Erdos-Renyi random graph (er-5000) to have the least significant clusters of all the graphs.

4.1 Distances between Clusterings of the Original and Perturbed Networks

Tables 2, 3, and 4 give the minimum, maximum, and mean (with standard devision) of the distance between the clusterings of the original network and of the 100 perturbed networks for all 9 combinations of clustering algorithm and distance measure. We point out that for each of the 11 networks the MCL algorithm always found that at least one of the perturbed networks had the same clusters as the original. As a result, the minimum distances in Table 4 are all 0.0.

Table 2 The minimum, average, and maximum distance computed between the clustering of the original network and the clustering of the 100 perturbed networks using the CNM algorithm

	Split-Join			Jaccard			NMI		
	Min	Mean (std)	Max	Min	Mean (std)	Max	Min	Mean (std)	Max
de2010	0.09	0.26 (.059)	0.36	0.28	0.54 (.081)	0.69	0.12	0.24 (.043)	0.35
pgp	0.00	0.07 (.034)	0.13	0.03	0.23 (.091)	0.40	0.00	0.06 (.027)	0.10
as-22july06	0.04	0.07 (.013)	0.09	0.13	0.31 (.059)	0.36	0.06	0.10 (.012)	0.12
power	0.08	0.11 (.013)	0.14	0.25	0.32 (.027)	0.37	0.06	0.08 (.0078)	0.10
del-n10	0.07	0.26 (.074)	0.37	0.18	0.53 (.012)	0.66	0.14	0.40 (.085)	0.50
del-n11	0.04	0.09 (.037)	0.20	0.14	0.26 (.086)	0.46	0.11	0.19 (.058)	0.33
del-n12	0.27	0.41 (.054)	0.47	0.57	0.70 (.041)	0.74	0.42	0.53 (.048)	0.59
gr30_30	0.07	0.11 (.035)	0.24	0.24	0.33 (.058)	0.53	0.20	0.28 (.045)	0.43
er-5000	0.62	0.67 (.011)	0.69	0.84	0.85 (.0026)	0.85	0.98	0.99 (.0030)	1.00
ba-10000	0.66	0.68 (.0067)	0.70	0.86	0.86 (.0014)	0.86	0.98	0.99 (.0016)	0.99
ws-10000	0.56	0.58 (.014)	0.62	0.85	0.86 (.0041)	0.87	0.74	0.75 (.019)	0.81

Both the CNM and Louvain algorithms produced significantly larger distances between clusterings for random graphs than for meshes or graphs from real-world networks, regardless of distance metric. In contrast, the MCL algorithm produced the largest distances for meshes, and distances for random graphs and real-world graphs were comparable. And in contrast to the CNM algorithm, both the Louvain and the MCL algorithms generally found larger distances between clusterings for meshes than for real-world networks. Finally, we observe that the Louvain algorithm finds much more stable clusters in the ws-10000 graph than the CNM algorithm. The magnitude of the difference is somewhat surprising since the Louvain algorithm is known to be better than the CNM algorithm at capturing clustering behavior at

Table 3 The minimum, average, and maximum distance computed between the clustering of the original network and the clustering of the 100 perturbed networks using the Louvain algorithm

	Split-Join			Jaccard			NMI		
	Min	Mean (std)	Max	Min	Mean (std)	Max	Min	Mean (std)	Max
de2010	0.21	0.25 (.020)	0.29	0.46	0.52 (.026)	0.58	0.12	0.14 (.0086)	0.16
pgp	0.06	0.11 (.021)	0.16	0.20	0.34 (.051)	0.45	0.05	0.08 (.013)	0.11
as-22july06	0.10	0.16 (.024)	0.23	0.26	0.39 (.055)	0.53	0.12	0.18 (.022)	0.23
power	0.11	0.17 (.023)	0.22	0.30	0.40 (.039)	0.47	0.08	0.11 (.013)	0.14
del-n10	0.24	0.32 (.037)	0.44	0.53	0.62 (.034)	0.70	0.22	0.27 (.021)	0.32
del-n11	0.16	0.30 (.036)	0.37	0.38	0.58 (.045)	0.66	0.15	0.23 (.021)	0.28
del-n12	0.28	0.34 (.027)	0.42	0.56	0.64 (.029)	0.71	0.22	0.26 (.014)	0.29
gr30_30	0.28	0.37 (.034)	0.45	0.60	0.67 (.027)	0.73	0.28	0.32 (.016)	0.35
er-5000	0.83	0.85 (.0063)	0.87	0.95	0.95 (.0025)	0.96	0.98	0.98 (.0022)	0.99
ba-10000	0.74	0.77 (.011)	0.79	0.89	0.91 (.0063)	0.93	0.99	0.99 (.0013)	1.00
ws-10000	0.07	0.12 (.019)	0.17	0.19	0.30 (.035)	0.38	0.04	0.06 (.0080)	0.08

Table 4 The minimum, average, and maximum distance computed between the clustering of the original network and the clustering of the 100 perturbed networks using the MCL algorithm

	Split-Join			Jaccard			NMI		
	Min	Mean (std)	Max	Min	Mean (std)	Max	Min	Mean (std)	Max
de2010	0.0	9.37e-05 (1.5e-04)	9.74e-04	0.0	3.50e-04 (6.0e-04)	3.14e-03	0.0	5.38e-05 (6.4e-05)	3.18e-04
pgp	0.0	1.35e-04 (1.7e-04)	9.36e-04	0.0	7.78e-04 (2.9e-03)	1.95e-02	0.0	8.15e-05 (1.0e-04)	5.40e-04
as-22july06	0.0	8.03e-05 (1.2e-04)	7.40e-04	0.0	2.23e-04 (3.3e-04)	1.40e-03	0.0	1.06e-04 (1.2e-04)	6.79e-04
power	0.0	7.36e-04 (6.2e-04)	3.04e-03	0.0	2.64e-03 (2.5e-03)	1.43e-02	0.0	3.54e-04 (2.8e-04)	1.37e-03
del-n10	0.0	6.05e-03 (9.7e-03)	6.10e-02	0.0	2.52e-02 (3.9e-02)	2.37e-01	0.0	9.21e-03 (1.0e-02)	6.10e-02
del-n11	0.0	5.52e-03 (6.2e-03)	2.91e-02	0.0	2.11e-02 (2.5e-02)	1.14e-01	0.0	5.98e-03 (5.2e-03)	2.30e-02
del-n12	0.0	1.73e-03 (2.2e-03)	1.03e-02	0.0	6.04e-03 (7.7e-03)	3.82e-02	0.0	1.57e-03 (1.5e-03)	6.17e-03
gr30_30	0.0	2.12e-03 (4.6e-03)	2.33e-02	0.0	8.15e-03 (1.7e-02)	8.64e-02	0.0	6.53e-03 (1.4e-02)	6.78e-02
er-5000	0.0	9.16e-04 (9.2e-04)	5.80e-03	0.0	4.86e-03 (6.9e-03)	5.62e-02	0.0	9.61e-04 (9.9e-04)	6.96e-03
ba-10000	0.0	1.41e-04 (1.7e-04)	9.00e-04	0.0	6.10e-04 (9.0e-04)	4.62e-03	0.0	4.56e-04 (5.6e-04)	2.95e-03
ws-10000	0.0	9.45e-05 (2.0e-04)	1.25e-03	0.0	3.52e-04 (7.1e-04)	3.84e-03	0.0	5.14e-05 (7.9e-05)	3.87e-04

different scales [3], but the Watts-Strogatz model, unlike the Barabasi-Albert model, does not generate scale-free graphs.

Of course, information is lost by looking only at the minimum, maximum, and mean of the distances between the clustering on the original graphs and the clusterings on the 100 perturbed graphs. The alternative would be to view plots of all the distances between clusterings, as in Figures 1 and 2. In these plots, each point represents a perturbed clustering. The vertical axis is distance between the original and perturbed clusterings, and the points are sorted horizontally by this distance. These figures show plots of the Split-Join, Jaccard, and NMI distances for clusterings computed using the CNM, Louvain, and MCL algorithms (in order from left to right) for the er-5000 (Figure 1) and pgp (Figure 2) networks. While the simple statistics highlight the difference in the values of the distances calculated, which are significantly greater for the synthetic er-5000 network than for the pgp network, the plots show that the distribution of the distances is also very different. This difference is most pronounced in clusters computed by the CNM algorithm, less so with the Louvain algorithm, and least with the MCL algorithm. Changing the measure of distance does not seem to affect the shape of the plot as much as the values.

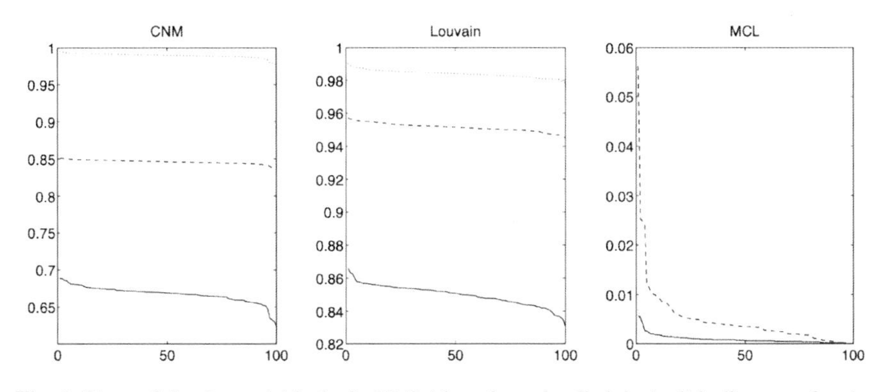

Fig. 1 Plots of the Jaccard (*dashed*), NMI (*dotted*), and split-join (*solid*) distances for the er-5000 network with clusters computed using the CNM (**left**), Louvain (**center**), and MCL (**right**) algorithms

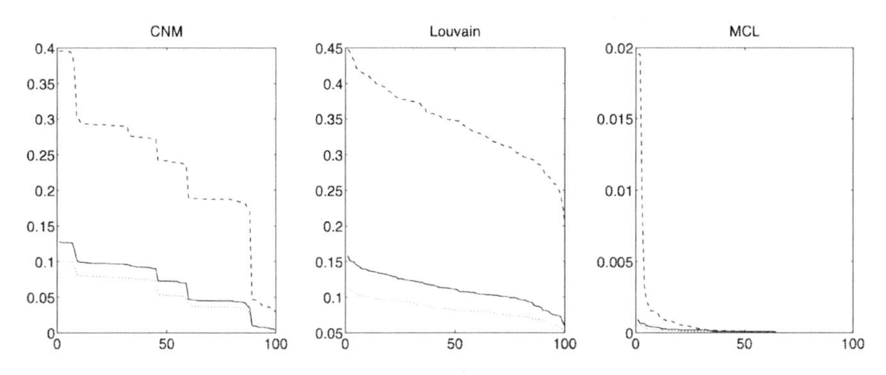

Fig. 2 Plots of the Jaccard (*dashed*), NMI (*dotted*), and split-join (*solid*) distances for the pgp network with clusters computed using the CNM (**left**), Louvain (**center**), and MCL (**right**) algorithms

4.2 *Modularity of Clusterings*

For each of the three clustering algorithms, Table 5 gives the modularity of the clustering of the original graph as well as the minimum and maximum modularity of the clusterings on all 100 perturbed networks.

For all three clustering algorithms the modularity values suggest that clusterings of random graphs are less significant than that of real-world graphs or of meshes.

Table 5 The modularity of the clusterings computed by the CNM, Louvain, and MCL algorithms

	CNM			Louvain			MCL		
	Min	Orig	Max	Min	Orig	Max	Min	Orig	Max
de2010	0.857	0.876	0.881	0.949	0.950	0.951	0.854	0.855	0.855
pgp	0.842	0.852	0.856	0.880	0.883	0.884	0.817	0.817	0.818
as-22july06	0.634	0.638	0.638	0.655	0.657	0.666	0.585	0.585	0.585
power	0.932	0.934	0.936	0.934	0.935	0.938	0.814	0.815	0.816
del-n10	0.706	0.727	0.737	0.805	0.810	0.817	0.795	0.798	0.807
del-n11	0.751	0.756	0.765	0.844	0.850	0.852	0.831	0.833	0.834
del-n12	0.760	0.769	0.781	0.874	0.877	0.879	0.846	0.848	0.849
gr30_30	0.679	0.695	0.703	0.772	0.778	0.781	0.698	0.699	0.699
er-5000	0.176	0.182	0.183	0.176	0.178	0.184	0.086	0.086	0.087
ba-10000	0.184	0.186	0.188	0.193	0.195	0.200	0.073	0.073	0.073
ws-10000	0.565	0.566	0.580	0.681	0.681	0.682	0.643	0.643	0.643

In addition, the modularity value suggests that the ws-10000 graph exhibits more significant clustering behavior than the other two random graphs. However, while Tables 2 and 3 show that on the ws-10000 graph the Louvain algorithm finds much more stable clusters than the CNM algorithm, Table 5 shows the resulting modularities are similar. This suggests that the modularity score alone obscures some information.

More generally, while the results in Tables 2, 3, and 4 show that a single edge can change the clustering, the results in Table 5 show that a single edge does not usually change the modularity by very much. This is evident, for example, with the gr30_30 matrix. This suggests that our strategy for evaluating clustering stability may also be able to provide information about what particularly influential edges could be added to an existing network.

5 Conclusion

In this work we describe a method for evaluating the stability of a clustering that is simpler and less expensive than those in the literature. We evaluate that method using three popular clustering algorithms and three measures of distance between clusters. We find that there are graphs on which individual clustering algorithms are known to give poor results, and that this can be seen in the stability of the clustering. We give evidence that this measure of stability provides information not available from just measuring modularity.

That said, this is not a complete study. In our experiments the network is always perturbed by adding an edge using the preferential attachment model. Several other models for adding edges are surveyed in [25]. Our experiments use only 3 of the many clustering algorithms that have been published and only 3 of the various distance measures that have been described in the literature. We did not experiment with parameters in the MCL clustering algorithms. Finally our methodology only ever perturbs a network by adding a single edge. Although this was sufficient to reveal interesting facts about the local stability of computed clusterings, incorporating larger perturbations might reveal similarly interesting facts about clustering behavior at a larger scale.

In addition to experimenting with some of the above factors, in the future we plan to look more carefully at the plots of the distances, rather than just at the minimum, average, and maximum values. It seems possible that the concavity of the plots may reveal information about the frequency of potential links with relatively large importance. We are also interested in generalizing our technique to weighted graphs and in experimenting with other methods of sampling edges. Finally, it would be valuable to investigate methods for evaluating the stability of subclusters. That is, even if there seems to be no global clustering behavior in a network, it is important to be able to identify smaller groups of vertices that exhibit tight clustering behavior.

References

1. Barabasi, A.L., Albert, R.: Emergence of scaling in random networks. Science 286(5439), 509–512 (1999), doi:10.1126/science.286.5439.509
2. Ben-Hur, A., Elisseeff, A., Guyon, I.: A stability based method for discovering structure in clustered data. In: Pac. Symp. on Biocomput., pp. 6–17 (2002), http://view.ncbi.nlm.nih.gov/pubmed/11928511
3. Blondel, V.D., Guillaume, J.L., Lambiotte, R., Lefebvre, E.: Fast unfolding of communities in large networks. J. of Stat. Mech.: Theory and Exp. 10, 8 (2008)
4. Blondel, V.D., Guillaume, J.L., Lambiotte, R., Lefebvre, E.: Louvain method: finding communities in large networks, https://sites.google.com/site/findcommunities/ (accessed: September 13, 2012)
5. Clauset, A.: "Fast Modularity" community structure inference algorithm, http://cs.unm.edu/aaron/research/fastmodularity.htm (accessed: August 15 , 2012)
6. Clauset, A., Newman, M.E.J., Moore, C.: Finding community structure in very large networks. Phys. Rev. E 70, 066111 (2004), doi:10.1103/PhysRevE.70.066111
7. Csardi, G., Nepusz, T.: The igraph software package for complex network research. Inter Journal Complex Systems 1695 (2006), http://igraph.sf.net
8. Danon, L., Díaz-Guilera, A., Duch, J., Arenas, A.: Comparing community structure identification. J. of Stat. Mech.: Theory and Exp. 2005 09,008 (2005)
9. Davis, T.: University of Florida sparse matrix collection. NA Digest, v.92, n.42, Oct. 16, 1994 and NA Digest, v.96, n.28, Jul. 23, 1996, and NA Digest, v.97, n.23 (June 7, 1997)
10. Dongen, S.V.: Graph clustering by flow simulation. Ph.D. thesis, University of Utrecht (2000)
11. Dongen, S.V.: Performance criteria for graph clustering and markov cluster experiments. Tech. Rep. INS-R0012, CWI (Centre for Mathematics and Computer Science) (2000)
12. Dongen, S.V.: MCL-edge: analyzing networks with millions of nodes, http://micans.org/mcl (accessed: August 20, 2012)
13. Erdos, P., Renyi, A.: On random graphs. Publ. Math (Debrecen) 6, 290 (1959)
14. Fortunato, S.: Community detection in graphs. Phys. Rep. 486, 75–174 (2010), doi:10.1016/j.physrep.2009.11.002

15. Gfeller, D., Chappelier, J.C., De Los Rios, P.: Finding instabilities in the community structure of complex networks. Phys. Rev. E 72, 056135 (2005), doi:10.1103/PhysRevE.72.056135
16. Good, B.H., de Montjoye, Y.A., Clauset, A.: Performance of modularity maximization in practical contexts. Phys. Rev. E 81, 046,106 (2010)
17. Guimera, R., Sales-Pardo, M., Amaral, L.: Modularity from fluctuations in random graphs and complex networks. Phys. Rev. E 70(2), 025,101 (2004)
18. Hagberg, A.A., Schult, D.A., Swart, P.J.: Exploring network structure, dynamics, and function using NetworkX. In: Proc. of the 7th Python in Sci. Conf. (SciPy 2008), Pasadena, CA USA, pp. 11–15 (2008)
19. Hu, Y., Nie, Y., Yang, H., Cheng, J., Fan, Y., Di, Z.: Measuring the significance of community structure in complex networks. Phys. Rev. E 82, 066106 (2010), doi:10.1103/PhysRevE.82.066106
20. Jaccard, P.: Étude comparative de la distribution florale dans une portion des Alpes et des Jura. Bull. del la Soc. Vaud. des Sci. Nat. 37, 547–579 (1901)
21. Karrer, B., Levina, E., Newman, M.E.J.: Robustness of community structure in networks. Phys. Rev. E 77, 046,119 (2008), doi:10.1103/PhysRevE.77.046119
22. Kleinberg, J.M.: An impossibility theorem for clustering. In: Becker, S., Thrun, S., Obermayer, K. (eds.) NIPS, pp. 446–453. MIT Press (2002)
23. Leskovec, J., Lang, K.J., Mahoney, M.: Empirical comparison of algorithms for network community detection. In: Proc. of the 19th Intl. Conf. on World Wide Web (WWW 2010), pp. 631–640. ACM, New York (2010), doi:10.1145/1772690.1772755
24. Levine, E., Domany, E.: Resampling method for unsupervised estimation of cluster validity. Neural Comput. 13(11), 2573–2593 (2001), doi:10.1162/089976601753196030
25. Liben-Nowell, D., Kleinberg, J.: The link prediction problem for social networks. In: Proc. of the Twelfth Intl. Conf. on Inf. and Knowl. Manag. (CIKM 2003), pp. 556–559. ACM, New York (2003), doi:10.1145/956863.956972
26. Meilă, M.: Comparing clusterings — an information based distance. J. of Multivar. Anal. 98(5), 873–895 (2007), doi:10.1016/j.jmva.2006.11.013
27. Mirshahvalad, A., Lindholm, J., Derlén, M., Rosvall, M.: Significant communities in large sparse networks. PLoS ONE 7(3), e33721 (2012), doi:10.1371/journal.pone.0033721
28. Moradi, F., Olovsson, T., Tsigas, P.: An Evaluation of Community Detection Algorithms on Large-Scale Email Traffic. In: Klasing, R. (ed.) SEA 2012. LNCS, vol. 7276, pp. 283–294. Springer, Heidelberg (2012)
29. Newman, M.E.J.: Fast algorithm for detecting community structure in networks. Phys. Rev. E 69, 066133 (2004), doi:10.1103/PhysRevE.69.066133
30. Newman, M.E.J.: Modularity and community structure in networks. Proc. of the Natl. Acad. of Sci. 103(23), 8577–8582 (2006), doi:10.1073/pnas.0601602103
31. Newman, M.E.J., Girvan, M.: Finding and evaluating community structure in networks. Phys. Rev. E 69, 026113 (2004)
32. Rosvall, M., Bergstrom, C.T.: Maps of random walks on complex networks reveal community structure. Proc. of the Natl. Acad. of Sci. 105(4), 1118–1123 (2008), doi:10.1073/pnas.0706851105
33. Rosvall, M., Bergstrom, C.T.: Mapping change in large networks. PLoS ONE 5(1), e8694 (2010), doi:10.1371/journal.pone.0008694
34. Vieira, V.: A Comparison of Methods for Community Detection in Large Scale Networks. In: Menezes, R., Evsukoff, A., González, M.C. (eds.) Complex Networks. SCI, vol. 424, pp. 75–86. Springer, Heidelberg (2013)
35. Watts, D.J., Strogatz, S.H.: Collective dynamics of 'small-world' networks. Nature 393(6684), 440–442 (1998), doi:10.1038/30918

Scalable Graph Clustering with Pregel

Bryan Perozzi, Christopher McCubbin, Spencer Beecher, and J.T. Halbert

Abstract. We outline a method for constructing in parallel a collection of local clusters for a massive distributed graph. For a given input set of (vertex, cluster size) tuples, we compute approximations of personal PageRank vectors in parallel using Pregel, and sweep the results using MapReduce. We show our method converges to the serial approximate PageRank, and perform an experiment that illustrates the speed up over the serial method. We also outline a random selection and deconfliction procedure to cluster a distributed graph, and perform experiments to determine the quality of clusterings returned.

1 Introduction

Recent developments in clustering algorithms have allowed the extraction of local clusters using a localized version of the PageRank algorithm known as "Personalized PageRank" [2]. Personal PageRank vectors can be efficiently computed by approximation. The "Approximate personal PageRank" (APR) approach concentrates the starting location of a normal PageRank in one vertex of the graph, and limits the distance that the PageRank walk and teleportation can progress.

One may then sort the surrounding vertices in decreasing order by their degree weighted probability from the APR vector, and then *sweep* them to search for the

Bryan Perozzi
Department of Computer Science, Stony Brook University, Stony Brook, NY, USA
e-mail: bperozzi@cs.stonybrook.edu

Christopher McCubbin
Sqrrl Data, Inc., Boston, MA, USA
e-mail: chris@sqrrl.com

Spencer Beecher · J.T. Halbert
TexelTek, Inc., Columbia, MD, USA
e-mail: {sbeecher, jhalbert}@texeltek.com

G. Ghoshal et al. (Eds.): *Complex Networks IV*, SCI 476, pp. 133–144.
DOI: 10.1007/978-3-642-36844-8_13

presence of a local cluster. This method generates a local good cluster if it exists, and runs in time proportionate to the size of the cluster.

In this work, we outline a method for constructing a collection of local clusters of a distributed graph in parallel. There are many possible applications for this technique, and we illustrate this with an example which uses local clusters to generate a clustering.

Our process to generate a clustering of a graph is composed of four steps that represent an extension of the work of Spielman and Teng [18] and Andersen et. al. [2].

The four steps are as follows.

1. We pick random pairs of source vertices and cluster sizes. The vertices are drawn by degree from the stationary distribution and the cluster sizes are drawn from a modification of Spielman's *RandomNibble* procedure. [18]
2. We compute the approximate personal PageRank vectors in parallel using Pregel [13] for each of these seed pairs.
3. We perform a sweep using MapReduce to produce the local clusters.
4. We reconcile cluster overlaps by assigning vertices to the cluster with lowest conductance. This is an implementation of an idea put forward by Andersen et. al. in an unpublished technical report.

Our contributions are the algorithms *Parallel Approximate PageRank* (PAPR) and *MapReduce Sweep*(MRSweep), which together can find local clusters in parallel on large graphs. We refer to their combination as *ParallelNibble*. We also provide proofs of convergence and asymptotic running time and experimental investigation of both the quality of clusterings produced and the algorithm's scalability on a variety of real world graphs.

2 Background

We will be considering an undirected graph $G = \{V, E\}$ where V is the set of vertices and $E \subseteq \{V \times V\}$ is the set of edges. Let $n = |V|$ and $m = |E|$.

As is commonly known many graphs exhibit community structure: that is, it is possible to group vertices into densely interconnected sets $C = \{C_i | C_i \subseteq V\}$. Extracting these communities from large graphs is something of an art since finding methods to evaluate the quality of a community is still an active area of research. We will discuss two of the more popular measures here.

One metric to describe the quality of a community C_i in a graph G is its *conductance*, $\phi(C_i)$. Intuitively $\phi(C_i)$ is the ratio between the perimeter of the cluster and its size [10]. It is defined as:

$$\phi(C_i) = \frac{|\text{outgoing edges of } C_i|}{\min(\text{Vol}(C_i), \text{Vol}(V \setminus C_i))}$$

Where $\text{Vol}(C_i) = \sum_{j \in C_i} \text{degree}(j)$. A lower conductance score therefore indicates a better cluster; the vertices are more tightly connected to each other than to vertices outside their community.

The conductance $\phi(C)$ of a clustering C in graph G is defined to be the minimum conductance of its clusters. This measure is favored by many authors (in particular Spielman et. al. in [18]) because of its connection to global spectral clustering via the Cheeger inequality. Additionally Andersen et. al. build a local version of the Cheeger inequality in [2].

Another metric to evaluate the quality of a clustering C is its modularity [16]. Modularity is designed to calculate the ratio of the internal edges of clusters in a given clustering to the number of edges that would be expected given a random assignment of edges. We calculate modularity with the following formula:

$$q(C) = \sum_{C_i \in C} \left\{ \frac{|E(C_i)|}{m} - \left(\frac{\sum_{v \in C_i} deg(v)}{2m} \right)^2 \right\}$$

Modularity satisfies the inequality $\frac{-1}{2} \leq q \leq 1$, and higher modularities are considered better clusterings.

3 Related Work

In this section we describe current publications in local graph algorithms, clustering, and distributed systems as it pertains to our work.

Graph Clustering Algorithms. An excellent overview of graph clustering is given in [17]. With the advent of so-called "big data", graphs with node and edge cardinalities in the billions or more have become common. This has created the need for algorithms that are scalable. Beginning with the famous PageRank algorithm, spectral methods for analyzing graphs have gained popularity in the past decade. Local spectral clustering methods, introduced by Speilman and Teng [18], and advanced by Andersen et. al. in [2] seek to apply these techniques scalably. The best time complexity to date comes from EvoCut [3], but does not beat the approximation guarantee of [2], which our work is an extension of.

An alternative to computing personal PageRank vectors in parallel is presented by Bahmani et. al. [5], who developed a fast Monte Carlo method for approximating a personal PageRank vector for every vertex in a graph using MapReduce. Our work uses a different approximation technique for personal PageRank vectors and is built on Pregel, but could perhaps be enhanced by their technique.

Local spectral methods and other local methods often require that a seed set of nodes is chosen. The problem of selecting the best starting vertices for local graph clustering has attracted some attention in the literature. Methods typically try to quickly compute metrics associated with good communities, and then use these results to seed community detection algorithms based on personalized PageRank. Recent work from Gleich and Seshadhri proposes a heuristic based on triangle counting in the vertex neighborhood [8].

A simple local graph clustering technique called *semi-clustering* was discussed in [13]. Our approach is computed in a different way, is optimizing a different cluster quality metric, and has different theoretical guarantees.

Nibble: An Algorithm for Local Community Detection. The Nibble algorithm was first sketched in [18] and more fully described in [19]. The algorithm finds a good cluster, of a specified size, near a given vertex. It runs nearly linearly in the size of the desired cluster, but is not guaranteed to succeed (i.e. such a set may not exist).

Nibble finds local clusters by computing an approximate distribution of a truncated random walk starting at the "seed" vertex. They extend the work of Lovász and Simonovits [12] to describe a set of conditions that find a low conductance cut based on these approximations quickly.

The PageRank-Nibble algorithm introduced by Andersen, Chung, Lang [2] improves upon this approach by using personal PageRank vectors to define nearness. They similarly extend the mixing result of Lovász and Simonovits to PageRank vectors.

Hadoop, MapReduce, and Giraph. Hadoop is an open-source implementation of the Distributed File System and MapReduce programming paradigm introduced in [7]. It has been used to implement a variety of algorithms for large graphs. [21,15]

Bulk Synchronous Parallel (BSP) processing is more flexible parallel computing framework than MapReduce. In BSP, a set of processors do computations. During these computations, the processors may send messages to other processors, either by name or as a broadcast. The computation will proceed until a barrier is reached in the algorithm. When the processor reaches a barrier, the system ensures that the processing will not continue until all the processors have reached the barrier. The system can then be seen as proceeding through a set of supersteps, marked by the barriers. Usually termination is done when all processors vote to halt at a barrier. If the virtual processors coincide with the nodes of a graph, we can perform many useful graph algorithms with BSP. This is the model used by the Pregel system [13] and later implemented in open source by the Apache Giraph project [4].

4 Algorithm

We use a three step process to compute a clustering of a graph. First, in PAPR, we compute many approximate personal PageRank vectors in parallel using the Pregel computing model. Next, in MRSweep, we perform a sweep of the vectors in parallel using Hadoop. These two algorithms together are a parallel version of the PageRank Nibble algorithm put forward in [2]. A critical difference of our *ParallelNibble* algorithm is that it produces local clusters which are overlapping. This prevents us from clustering graphs with a straightforward application of the Partition algorithm from [18]. The final step of our process transforms these overlapping local clusters into a non-overlapping clustering of the graph.

Parallel Approximate PageRank. The computation of full PageRank has been intimately associated with the Pregel framework [14]. Here we present the approach for computing an approximate personal PageRank vector in Pregel, an outline of the proof of correctness showing that approximate personal PageRank vectors computed in this way still converge to personal PageRank vectors, and an analysis of the amount of work required to perform this computation.

Computing an approximate personal PageRank vector. We compute the approximate personal PageRank vector with a direct parallelization of the approach of Andersen, et. al. [2]. We start with a PageRank vector p of all zeros and a residual vector r initially set to $r = \chi_v$, i.e. the vector of all zeros except a 1 corresponding to the source vertex v.

The two inputs to this algorithm are $\{v\}$, the set of source nodes, and b, the log volume of the desired cluster. In the following proofs we follow the notation of [2] which uses the energy constant $\varepsilon \in [0,1]$, and the teleportation constant $\alpha \in [0,1]$. The energy constant ε controls the approximation level of the personal PageRank vector. As the size of the desired cluster grows, a finer approximation is necessary. In practice, we require $\varepsilon = O(\frac{1}{2^b})$ and typically initialize $\alpha = 0.10$.

Algorithm 1. PAPR(v,ε,α)

At each vertex u, for each superstep:

1. If this vertex has any messages i from a neighbor pushing weight w_i from the last step, set $r_u = r_u + w_i$
2. If $\frac{r_u}{d_u} > \varepsilon$ perform the push operation at this vertex.
3. If $\frac{r_u}{d_u} < \varepsilon$, vote to halt.

We define the push operation at a vertex to be:

1. set $p_u = p_u + \alpha r_u$
2. set $r_u = (1-\alpha)r_u/2$
3. for each of my neighbors, send a message with weight $w = (1-\alpha)r_u/2d_u$ attached.

In Pregel, each vertex has a corresponding processor and state. We realize the vectors p and r in our implementation by storing the scalar associated with vertex i in its processor. Along with p_i and r_i, we also store the values of the global parameters m, α, and ε. The algorithm for each vertex in a superstep is given in Algorithm 1.

Run this algorithm until all nodes vote to halt. We will show that PAPR halts and converges to an approximate PageRank vector; the number of push operations performed by PAPR is $O(\frac{1}{\varepsilon\alpha})$; and the complexity of PAPR is $O(\frac{1}{\varepsilon\alpha\omega})$ where ω is the number of workers. The basic ideas of these proofs follow in spirit along with proofs in [2], except where one step at at time is considered in those proofs, multiple steps may be performed in parallel by our algorithm. We can show that equivalent steps will be performed at each vertex as in the original algorithm up to a reordering, and therefore the same results hold due to the linearity of the functions involved.

Proof that PAPR Terminates and Converges to Approximate PageRank

Lemma 1. Let U be the set of nodes that experience the push operation in a superstep. After the push operation, our algorithm will produce the vectors:

$$p' = p + \sum_{u \in U} \alpha r(u) \chi_u \tag{1}$$

$$r' = r - \sum_{u \in U} \{ r(u)\chi_u + (1-\alpha) r(u) \chi_u W \} \tag{2}$$

Proof:

It is evident that p' is of the form described by the definition of the algorithm.

We can simplify the equation for r' to:

$$r - \sum_{u \in U} \left\{ r(u)\chi_u + \frac{1}{2}(1-\alpha) r(u) \left(\chi_u + \frac{\chi_u A}{d(u)} \right) \right\} \tag{3}$$

Using this simplification, we can compare components with what the algorithm will produce. If an element v of r corresponds to a vertex that is not in U or U's neighbors, then all the components in equation 3 besides the first are 0, so $r'(v) = r(v)$ like we expect. Otherwise, if v is not in U but is a neighbor of U, equation 3 has as the components of $r(v)$

$$r(v) - \sum_{u \in U} \left\{ \cancelto{0}{r(u)\chi_u(v)} + \frac{1}{2}(1-\alpha) r(u) \left(\cancelto{0}{\chi_u(v)} + \frac{\chi_u A(v)}{d(u)} \right) \right\}$$

$$r(v) - \sum_{(v,u) \in E} \left\{ \frac{1}{2}(1-\alpha) r(u) \frac{1}{d(u)} \right\}$$

Which is what you would expect. When v is in U, the χ_u factors cancel to 1 when $v = u$ so we get

$$\cancelto{0}{r(v) - r(v)} + \frac{r(v)(1-\alpha)}{2} + \sum_{(v,u) \in E} \frac{(1-\alpha) r(u)}{2d(u)}$$

which is also what we expect, proving the lemma.

Lemma 2. To show that PAPR converges to APR, we need to show that in PAPR as in APR, $p + pr(\alpha, r) = p' + pr(\alpha, r')$.

Using equation 5 in [2] and the linearity of the pr function,

$$\begin{aligned} p + pr(\alpha, r) &= p + pr(\alpha, r - \sum_{u \in U} r(u)\chi_u) + \sum_{u \in U} pr(\alpha, r(u)\chi_u) \\ &= p + pr(\alpha, r - \sum_{u \in U} r(u)\chi_u) + \sum_{u \in U} [\alpha r(u)\chi_u + (1-\alpha) pr(\alpha, r(u)\chi_u W)] \end{aligned}$$

$$
\begin{aligned}
&= p' + pr(\alpha, r - \sum_{u \in U} [r(u)\chi_u) + (1-\alpha)r(u)\chi_u W]) \\
&= p' + pr(\alpha, r')
\end{aligned}
$$

Lemma 3. Let T be the total number of push operations performed by Parallel Approximate PageRank, S be the number of supersteps, ω be the number of workers and d_i be the degree of the vertex u used in the ith push. We would like to show that $\sum_{i=1}^{T} d_i \leq \frac{1}{\varepsilon\alpha}$

Proof: The proof follows as the proof in [2]. However, in PAPR many push operations are performed in each superstep. We can number the push operations using an index i, using the constraint that a push operation in an earlier superstep than another always has a lower index (numbering within a superstep is arbitrary). Since we use the same condition to choose vertices to perform the push operation on as in [2], each individual push operation on a vertex taken by itself still decreases $|r|_1$ by an amount greater than $\varepsilon\alpha d_i$. The result follows.

Complexity of PAPR. Consider S, the vector of super step lengths.

We partition the algorithm into $|S|$ super steps, such that $\sum_{S_i \in S} S_i = T$, i.e. S_i represents the number of pushes in step i. So then

$$
\sum_{i=0}^{T} d_i = \sum_{S_j \in S} \sum_{i=0}^{S_j} d_i \implies \varepsilon\alpha \sum_{i=0}^{T} d_i = \sum_{S_j \in S} \varepsilon\alpha \sum_{i=0}^{S_j} d_i
$$

Consider ω workers, each of which have been assigned $\frac{|Supp(p)|}{\omega}$ i.i.d. vertices for computation in parallel (i.e. the vertices with non-zero entries in p are divided uniformly among ω). We can then write the total amount of work in terms of the expected amount of work performed by each worker per superstep:

$$
\varepsilon\alpha \sum_{i=0}^{T} d_i = \sum_{S_j \in S} \varepsilon\alpha \sum_{i=0}^{S_j} d_i = \sum_{S_j \in S} \varepsilon\alpha\omega \sum_{i=0}^{S_j/\omega} d_i
$$

This implies $\varepsilon\alpha\omega \sum_{S_j \in S} \sum_{i=0}^{S_j/\omega} d_i = \varepsilon\alpha \sum_{i=0}^{T} d_i \leq 1$, as in the proof by [2], because $||r||_1 = 1$. This would them imply that the total running time satisfies the relationship $\sum_{S_j \in S} \sum_{i=0}^{S_j/\omega} d_i \leq \frac{1}{\varepsilon\alpha\omega}$. Therefore PAPR's complexity is $O(\frac{1}{\alpha\varepsilon\omega})$.

Computing multiple APRs simultaneously. In the previous section, we showed that we can compute one APR from a starting vertex v using a parallel algorithm. To compute more APRs from a set of starting points S, we simply store a scalar pagerank entry p_j and residual entry r_j for each starting vertex $s_j \in S$, and initialize appropriately. We then modify the algorithm to compute each scalar quantity in turn for each starting vertex.

MapReduce Sweep. In [2], each APR vector is converted into a good clustering using a *sweeping* technique. One orders the nodes in the graph using the corresponding probability value in the personal PageRank vector divided by the degree: $\frac{p_n}{d_n}$. If the PageRank vector has a support size equal to a number N_p, this creates an ordering on the nodes $n_1, n_2, \ldots, n_{N_p}$ and induces *sweep sets* $S_j^p = \{n_i | i \leq j\}$. A set with good conductance is found by finding the set with minimum conductance out of these sweep sets, but will output nothing if the set's conductance is greater than ϕ_{min}.

In the graphs that we are considering, we wish to compute many such good sets in parallel and also leverage the power of the MapReduce framework to aid in the algorithm computation. Between the Map and Reduce phases of MapReduce, the keys emitted by the Mapper are both partitioned into separate sets, and within each partition the keys are sorted according to some comparator. Keys present in a partition are guaranteed to be processed by the same Reducer, in sorted order.

In our MapReduce implementation of the Sweep algorithm, the Mapper will iterate over the vertices output by the Pregel APR algorithm. This output contains the probability value for each APR vector computed that affected that vertex, as well as the vertex's degree. We create keys emitted by the Mapper that are partitioned by the APR start vertex, and are sorted by the sweep metric $\frac{p_n}{d_n}$. Therefore after the Map phase the Reducers will receive all the probability and degree values for a single APR vector, sorted the correct way to produce the sweep sets. The Reducer then can compute the conductance for each APR's sweep sets and find the minimum conductance value. As an additional optimization, the data structure used to compute prior conductance values can be re-used to quickly compute the conductance value for the same set with an additional vertex. One simply stores the structures needed to compute conductance, such as the set of vertices adjacent to the cluster but not in the cluster, and updates them as new vertices arrive with their neighbor data.

4.1 Clustering Whole Graphs

The ParallelNibble procedure presented above provides a way of computing local clusters in parallel on distributed graphs. The ability to detect local communities is useful in a variety of real-world graph analysis tasks when one wants to know more about a source node (e.g. in a social network such as Twitter, one could model a node's local community affiliation and use it to determine interest in trending topics.)

To further explore the power of these local methods, we now consider the problem of generating a clustering for an entire graph. In order to do this we require an approach to generate good candidate tuples of source nodes and cluster sizes to build local clusters from, and a method for dealing with overlapping clusters.

Selection of source vertices and cluster sizes. As with all local clustering methods, the selection of the starting vertices will make a significant difference in the final clustering. To generate tuples (v_i, b_i) as input for ParallelNibble we take inspiration from Spielman and Teng's RandomNibble [18]. Specifically, for each desired candidate nibble i, we randomly select a vertex v_i from the stationary distribution and a cluster size b_i in the range $\frac{\lceil \log m \rceil}{2}$, ..., $\lceil \log m \rceil$ according to $\Pr[b = i] = 2^{-i}/(1 - 2^{-\lceil \log m \rceil})$.

This distribution for b is a truncated version of RandomNibble's; it focuses on finding larger clusters instead of smaller ones. Choosing b this way makes sense for performing a coarse clustering of G, but it does have a disadvantage - this approach will be unable to detect small clusters. A remedy for this is to recursively apply the same procedure to the generated clusters.

Postprocessing overlapping clusters. Once we have computed the local clusters for all the source vertices, we wish to convert them into a good global clustering. There are a variety of ways that these overlapping local clusters could be combined. We choose a simple method put forward by Andersen, et. al. which has the desirable property of preserving the minimum conductance of the final clustering. The method amounts to resolving conflicts in local cluster membership by always assigning a vertex to its cluster with the least conductance. It is accomplished by the following procedure.

First, we sort the generated clusters by their conductance. Then we iterate through the clusters, adding them to our final clustering. As we add each cluster the final clustering, we mark all of the vertices in it as 'used'. Clusters with higher (worse) conductances can not use these vertices again.

This is clearly not optimal for maximizing the modularity of the clustering, but provides a straightforward approach for dealing with a complicated problem.

5 Experimental Results

Here we present results obtained from running our algorithm on real graphs. We focus on two types of metrics: the quality of the clustering in terms of conductance or modularity, and the algorithm scalability measured by the running time vs. number of worker processes.

Test Environment. We used two different environments, a 12 machine cluster for the quality of clustering tests, and a 32 machine cluster for the scalability test. Each machine has 96 GB of RAM, and 2 Intel Xeon processors. The cluster was using Apache Hadoop 0.20.203 on CentOS 6. All experiments were written in Scala, using Apache Giraph 0.2.

Evaluation of Clustering. In order to evaluate the quality of the clustering found by our algorithm, we have benchmarked it against a variety of real world graphs found from the SNAP Network Datasets[1]. We compare our results against *Louvain Fast Unfolding* [6] a popular modularity optimization algorithm that performs a local search followed by a cluster contraction phase which repeats until it finds a maximum modularity. Fast Unfolding is not optimal, but it is quite fast and has been shown to achieve excellent results on a variety of real world graphs.

We emphasize that Fast Unfolding is an algorithm optimizing a *global* criteria (modularity) using local changes, while Nibble and its derivatives are completely local algorithms optimizing a local criteria (the conductance of *local cuts*). In some

[1] Available: `http://snap.stanford.edu/data/index.html`

cases, Fast Unfolding is barely able to run and must be supplied with multiple gigabytes of memory. We compare the results of our clustering process against the first phase of Fast Unfolding before cluster contraction is applied. We refer to this as the *Baseline modularity*. The conductance of this clustering we use as our *baseline conductance*.

For all these tests, we used the number of Giraph workers $\omega = 10$, the teleportation constant $\alpha = 0.10$, and the minimum acceptable conductance of MRSweep $\phi_{min} = 0.15$.

Table 1 shows that PAPR is able to find low conductance clusters, but that the complete clustering performs worse than the baseline modularity. Potential ways to improve performance include taking more samples, changing the way sources are selected, or calculating a more precise PageRank vector.

Table 1 The quality of clusterings produced by our method on some publicly available graph datasets. We have ignored clusters with $\phi = 0$ (this indicates a disconnected cluster was discovered). Baseline modularity (q) and baseline ϕ are derived from a clustering made with the Fast Unfolding method [6]

Graph Name	$\|V\|$	$\|E\|$	ϕ	Baseline ϕ	q	Baseline q
soc-livejournal	4,847,571	68,993,773	0.0376	0.1764	0.488	0.665527
web-google	875,713	5,105,039	0.027	0.015	0.689	0.76056
web-stanford	281,903	2,312,497	0.017	0.001	0.584	0.815849
amazon0302	262,111	1,234,877	0.109	0.03846	0.617	0.637707
ca-AstroPh	18,772	396,160	0.1507	0.0666	0.244	0.54332
ca-GrQc	5,242	28,980	0.106	0.004	0.538	0.708325
ca-HepPh	12,008	237,010	0.120	0.002	0.473	0.587588
email-Enron	36,692	367,662	0.066	0.15	0.384	0.557363
loc-gowalla	196,591	950,327	0.107	0.1515	0.459	0.639371
oregon1-010526	10,670	22,002	0.145	0.1594	0.438	0.458858
soc-Epinions1	75,879	508,837	0.164	0.2	0.271	0.405964
web-Stanford	281,903	2,312,497	0.022	0.001	0.54	0.81631
wiki-Vote	7,115	103,689	0.178	0.111	0.295	0.42207

PAPR Scalability. To verify the scalability of our local clustering approach we computed a fixed number of clusters on the biggest graph we considered, *soc-livejournal*, and varied the number of workers, ω, available to the Apache Giraph job. Source vertices were selected randomly, but the cluster size was fixed. Other parameters are as used earlier. The total time to run the algorithm includes the time for Giraph to load the graph, and the time to run PAPR. We present the total time and the PAPR running time in Figure 1.

As expected, increasing the number of workers decreases the running time. After a certain number of workers (ω=10 in this case) the synchronization and communication costs begin to dominate the computation, and there is no benefit from additional parallelization.

One of the prime contributors to this communication overhead comes from the difficulty of partitioning graphs which follow a power-law distribution. When faced with such a graph Pregel randomly assigns the vertices to workers. This results in most of the edges running between different workers and requires network overhead for messages passed over these edges. Recent work by Gonzalez et. al. [9] presents the problem in detail and provides a computational approach using vertex-cuts instead of edge-cuts which allows for much greater parallelization.

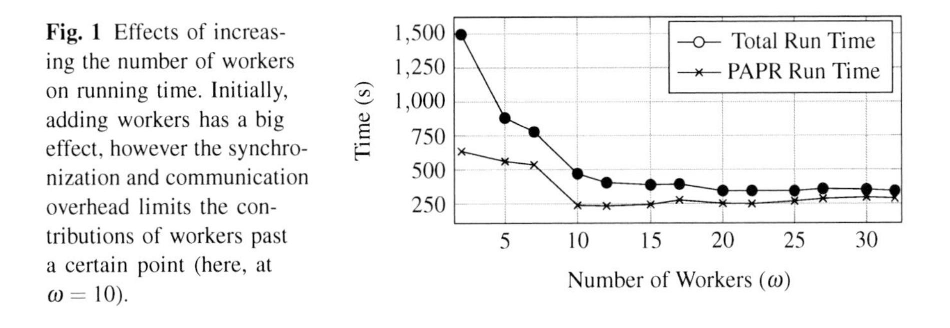

Fig. 1 Effects of increasing the number of workers on running time. Initially, adding workers has a big effect, however the synchronization and communication overhead limits the contributions of workers past a certain point (here, at $\omega = 10$).

6 Conclusions and Future Work

We have shown that a parallel technique can be used to create Approximate PageRank vectors and turn those vectors into local clusters using the parallel processing techniques of Pregel and MapReduce. We have shown that the calculation of these vectors is highly parallelizable and results in significant time savings as workers are added. This time and memory parallelization allows the use of these local spectral clustering techniques on larger graphs than would traditionally be possible by simply adding more commodity hardware to the analysis system.

Recent work on community detection [1, 20] shows that allowing communities to overlap better captures the behavior observed in in real world networks. Methods based on local clustering have already been used to analyze the profile of network communities at different size scales [11], and there is reason to believe that these techniques can aide in other aspects of the analysis of large graphs. We plan to perform more scalability analysis of the technique using more hardware and "Internet scale" graphs reaching into the billions of nodes, where traditional methods have serious difficulty providing meaningful results.

References

1. Ahn, Y.Y., Bagrow, J.P., Lehmann, S.: Link communities reveal multiscale complexity in networks. Nature 466(7307), 761–764 (2010)
2. Andersen, R., Chung, F., Lang, K.: Local graph partitioning using pagerank vectors. In: 47th Annual IEEE Symposium on Foundations of Computer Science, FOCS 2006, pp. 475–486. IEEE (2006)

3. Andersen, R., Peres, Y.: Finding sparse cuts locally using evolving sets, CoRR, abs/0811.3779 (2008)
4. Apache giraph (February 2012), http://incubator.apache.org/giraph/
5. Bahmani, B., Chakrabarti, K., Xin, D.: Fast personalized pagerank on mapreduce. In: Proceedings of the 2011 ACM SIGMOD International Conference on Management of Data, SIGMOD 2011, pp. 973–984. ACM, New York (2011)
6. Blondel, V.D., Guillaume, J.L., Lambiotte, R., Mech, E.L.J.S.: Fast unfolding of communities in large networks. J. Stat. Mech., P10008 (2008)
7. Dean, J., Ghemawat, S.: Mapreduce: simplified data processing on large clusters. Commun. ACM 51(1), 107–113 (2008)
8. Gleich, D.F., Seshadhri, C.: Vertex neighborhoods, low conductance cuts, and good seeds for local community methods. In: KDD, pp. 597–605 (2012)
9. Gonzalez, J.E., Low, Y., Gu, H., Bickson, D., Guestrin, C.: Powergraph: distributed graph-parallel computation on natural graphs. In: Proceedings of the 10th USENIX Conference on Operating Systems Design and Implementation, OSDI 2012, pp. 17–30. USENIX Association, Berkeley (2012)
10. Kannan, R., Vempala, S., Vetta, A.: On clusterings: Good, bad and spectral. J. ACM 51(3), 497–515 (2004)
11. Leskovec, J., Lang, K.J., Dasgupta, A., Mahoney, M.W.: Statistical properties of community structure in large social and information networks. In: roceedings of the 17th International Conference on World Wide Web, WWW 2008, pp. 695–704. ACM, New York (2008)
12. Lovász, L., Simonovits, M.: Random walks in a convex body and an improved volume algorithm. Random Structures & Algorithms 4(4), 359–412 (1993)
13. Malewicz, G., Austern, M.H., Bik, A.J.C., Dehnert, J.C., Horn, I., Leiser, N., Czajkowski, G.: Pregel: a system for large-scale graph processing. In: Proceedings of the 2010 ACM SIGMOD International Conference on Management of Data, SIGMOD 2010, pp. 135–146. ACM, New York (2010)
14. Malewicz, G., Austern, M.H., Bik, A.J.C., Dehnert, J.C., Horn, I., Leiser, N., Czajkowski, G.: Pregel: a system for large-scale graph processing. In: Proceedings of the 2010 ACM SIGMOD International Conference on Management of Data, SIGMOD 2010, pp. 135–146. ACM, New York (2010)
15. McCubbin, C., Perozzi, B., Levine, A., Rahman, A.: Finding the 'needle': Locating interesting nodes using the k-shortest paths algorithm in mapreduce. In: 2011 IEEE International Conference on Data Mining Workshops, pp. 180–187 (2011)
16. Newman, M.E.J.: Modularity and community structure in networks. Proceedings of the National Academy of Sciences 103(23), 8577–8582 (2006)
17. Schaeffer, S.E.: Graph clustering. Computer Science Review 1(1), 27–64 (2007)
18. Spielman, D.A., Teng, S.H.: Nearly-linear time algorithms for graph partitioning, graph sparsification, and solving linear systems. In: Proceedings of the Thirty-Sixth Annual ACM Symposium on Theory of Computing, pp. 81–90. ACM (2004)
19. Spielman, D.A., Teng, S.-H.: A local clustering algorithm for massive graphs and its application to nearly-linear time graph partitioning, CoRR, abs/0809.3232 (2008)
20. Yang, J., Leskovec, J.: Structure and overlaps of communities in networks, CoRR, abs/1205.6228 (2012)
21. Zhao, Z., Wang, G., Butt, A.R., Khan, M., Kumar, V.S., Marathe, M.V.: Sahad: Subgraph analysis in massive networks using hadoop. In: 2012 IEEE 26th International Parallel & Distributed Processing Symposium (IPDPS), pp. 390–401. IEEE (2012)

Unfolding Ego-Centered Community Structures with "A Similarity Approach"

Maximilien Danisch, Jean-Loup Guillaume, and Bénédicte Le Grand

Abstract. We propose a framework to unfold the ego-centered community structure of a given node in a network. The framework is not based on the optimization of a quality function, but on the study of the irregularity of the decrease of a similarity measure. It is a practical use of the notion of multi-ego-centered community and we validate the pertinence of the approach on a real-world network of wikipedia pages.

1 Context and Related Work

Many real-world complex systems, such as social or computer networks can be modeled as large graphs, called complex networks. Because of the increasing volume of data and the need to understand such huge systems, complex networks have been extensively studied these last ten years. Due to its applications, notably in market research and classification, and its intriguing nature, the notion of communities of nodes[1] and their detection has been at the center of this research. For an extensive survey on community detection, we refer to [FOR10].

Communities are clearly overlapping in real world systems, especially in social networks, where every individual belongs to various communities: family, colleagues, groups of friends, etc. Finding all these overlapping communities in a huge graph is very complex: in a graph of n nodes there are 2^n such possible communities and 2^{2^n} such possible community structures. Even if these communities could be efficiently computed, it may lead to uninterpretable

Maximilien Danisch · Jean-Loup Guillaume
LIP6, Université Pierre et Marie Curie, 4 Place Jussieu, 75005 Paris, France

Bénédicte Le Grand
CRI, Université Paris 1 Panthéon-Sorbonne. 90 rue de Tolbiac, 75013 Paris, France

[1] Groups of nodes very connected to one-another, but loosely connected to the outside.

G. Ghoshal et al. (Eds.): *Complex Networks IV*, SCI 476, pp. 145–153.
DOI: 10.1007/978-3-642-36844-8_14

results. However, some studies have still tackled this problem, such as [PAL05] and [EVA09].

Because of the complexity of overlapping communities detection, most studies have restricted the community structure to a partition, where each node belongs to one and only one community. This problem, also very complex, does not have a perfect solution for now, however several algorithms with very satisfying results exist, in particular the Louvain method [BLO08] which optimizes the *modularity* [GIR02] in an agglomerative fashion, and Infomap [ROS08].

Another approach, to keep the realism of overlapping communities, but without making the problem too complex, is to focus on a single node and try to find all the communities it belongs to, which we call *ego-centered communities*. This has been extensively studied following a quality function approach: starting from a group where only the given node is included and optimizing step by step (by adding or removing a node from the group) a given quality function, see [CLA05, CHE09, FRI11, NGO12].

However this quality function approach suffers from two important drawbacks: (i) designing a good cost function is very difficult, particularly because of a problem of hidden scale parameters. For instance in [FRI11], the quality function, *cohesion*, incorporates a term measuring the density of triangle, which decreases in $O(s^3)$ (where s is the number of selected nodes) in sparse graphs. This thus leads to very small communities in sparse graphs. This problem could be coped by decreasing the effect of this density term, for instance by taking its power a ($a \leq 1$), which is a hidden scale parameter set to one in *cohesion*. (ii) Optimizing the quality function is also very hard because of the highly non-convex nature of the optimization landscape, which leads again to small communities. Indeed, as the optimization is conducted in a greedy way (any other method leading to very slow algorithms), it is thus missing large communities if the algorithms needs to go through lower values of the quality function to reach higher values corresponding to large communities.

In this article we propose a transversal approach to find ego-centered communities of a given node which we will detail next. We show the result of our method when applied to a real large graph: the whole wikipedia network containing more than 2 million labeled pages and 40 million edges hyperlinks [PAL08].

2 Framework

Given a specific node u, we measure the similarity[2] of all nodes in the graph to u and then try to find irregularities in the decrease of these similarity values, instead of optimizing a quality function. Such irregularities can reflect the

[2] Even though other similarity measures can be used, we use the carryover opinion introduced in [DAN12].

presence of one or more communities. More precisely, if there exists a group of nodes that are equally similar to the node of interest, while all other nodes are less similar to it, then sorting in decreasing order and plotting these similarity values will lead to two plateaus separated by a strong decrease. The nodes before the strong decrease constitute a community of u. However this routine often leads to a power-law with no plateau and from which no scale can be extracted; this happens when lots of communities of various sizes are overlapping which is often the case. To cope with this problem, we use the notion of multi-ego-centered community, i.e., centered on a set of nodes instead of a single node. The key idea is that, although one node generally belongs to numerous communities, a small set of appropriate nodes can fully characterize a single community.

We thus need to smartly pick another node, v, evaluate the similarity of all nodes in the graph to v and then for each nodes in the graph, compute the minimum of the score obtained from u and the score obtained from v: this minimum evaluates how a node is similar to u AND v. Once again, if there exists a group of nodes that are equally similar to u AND v, while all other nodes are less similar to it, then this group can constitute a community. Note that doing this sometimes leads to the identification of a community which does not contain u and/or v, however since we are interested only in communities containing u, we use v as an artifact and keep a community only if it contains u, regardless of v. The framework consists in doing this for enough candidate nodes v in order to obtain all communities of u. We will now detail the steps of the framework.

2.1 How to Chose the Candidates for v?

First, the Carryover opinion of u has to be computed [DAN12]. This gives a real value for each node present in the graph: its similarity to u. Sorting the obtained values and plotting them as a function of their ranking leads to the carryover curve. If the outcome is a power-law, there is no relevant scale and u certainly belongs to several communities of various sizes.

We then want to pick v such that v and u roughly share exactly one community. If v is very dissimilar from u then it is very unlikely that u and v will share a common community: computing the minimum of the scores obtained from running the carryover opinion from u and the scores obtained from running the carryover opinion from v will lead to very small values. Indeed if the two nodes share no community, at least one of the scores will be very low. Conversely if v is extremely similar to u then the two nodes will share many communities. The carryover opinion values obtained from u and v will be roughly the same and doing the minimum will not give more information.

Thus v must be similar enough to u, but not too similar: it has to have a carryover score obtained from u not too high and not too low. A low and high similarity thresholds can be manually tuned to select all nodes at the right distance in order to fasten the execution.

It is quite likely that many of these nodes at the right distance will lead to the identification of the same community, therefore not all of them need to be candidates; a random selection of them can be used if the running time of the algorithm matters. More precise selection strategies will be discussed in the future work section.

2.2 *How to Identify the Ego-Centered Community of u and v?*

In order to identify the potential community centered on both u and v, we must compute the minimum of the carryover values obtained from u and from v for each node, w, of the graph. The minimum of the two scores is therefore a measure of the belonging of w to the community of both u and v. We can then sort these minimum values and plot the minimum carryover curve. As before, an irregularity in the decrease, i.e., a plateau followed by a strong decrease, indicates that all nodes before the decrease constitute a community.

Detecting a plateau followed by a strong decrease can be done automatically: if the maximum slope is higher than a given threshold, the nodes before this maximum slope constitute a community. This threshold should be manually tuned. If there are several sharp decreases, we only detect the sharpest, this could be improved in the future.

In addition, if u is before the decrease then u is in the community. In that case, these nodes before the decrease constitute a community of u. Note that v does not need to belong to this community since we are trying to find communities around u and that v is only a node that we use to find such communities.

As such this method is not very efficient when the carryover opinion is run from a very high degree node connected to a very large number of communities. In that case, the carryover tends to give high values to every node in the graph and calculating the minimum with the scores obtained from a less popular node, which gives lower values to the nodes, will simply result in the values obtained with this second node. A rescaling before doing the minimum can fix the problem. Indeed the lowest values obtained by running the carryover opinion result in a plateau, rescaling (in logarithmic scale) the values such that these plateaus are at the same level solves this problem.

2.3 *Cleaning the Output and Labeling the Communities*

The output of the two previous steps is a set of communities (where each node is scored), since each candidate node can yield a community. These communities need to be postprocessed, since many of them are very similar.

We propose to clean the output as follows: if the Jaccard similarity [3] between two communities (or any other similarity measure between sets) is too large, it means that although communities are actually the same, they appear to be different because of the noise. In that case we only keep the intersection of these two communities. For each node in this new (intersection) community, the score is given by the sum of the scores in the original communities.

We perform an optional cleaning step, which enhances the results: if a community is dissimilar to all other communities, we simply remove it. Indeed, a good community should appear for different candidate nodes. We observed that such communities come from the detection of a plateau/decrease structure which does not exist (it often happens when the threshold is not set to a proper value).

We finally label the community with the label of the best ranked node in the community, i.e., the node whose sum of values is the highest. If two communities have the same label we suggest to keep both (it can be different scales of the same community).

This algorithm finally gives a set of distincts labeled communities. We now show some results on a real network.

3 Results and Validation

Because of size limitation, we focus here on the result for a single node, the wikipedia page entitled *Chess Boxing* [4]. This page exhibits good results which are easily interpretable and can be validated by hand.

For the "Chess Boxing" node, the algorithm iterated over 3000 nodes chosen at random from the nodes between the 100th and the 10.000^{th} best ranked nodes leads to 770 groups of nodes. Figure 1 shows a successful trial leading to the identification of a group along with an unsuccessful trial.

Figures 2a shows the Jaccard similarity matrix of the 770 unfolded communities before cleaning. The columns and lines of the matrix have been rearranged so that columns corresponding to similar groups are next to each other. We see that there are 716 communities very similar to one another, while not similar to the other ones. If u is in or around a large community, we have a high probability to unfold it, and this probability increases with the size of the community. A problem of the algorithm is that if very large communities exist, the algorithm can have some difficulty to unfold other small communities. We will come back to that problem in the future work section.

When zooming on the rest of the matrix, figure 2b, we see 4 medium size groups of communities and 6 groups containing only a single community, these

[3] For two sets A and B, the Jaccard similarity is given by $Jac(A, B) = \frac{|A \cap B|}{|A \cup B|}$.

[4] ChessBoxing is a sport mixing Chess and Boxing in alternated rounds.

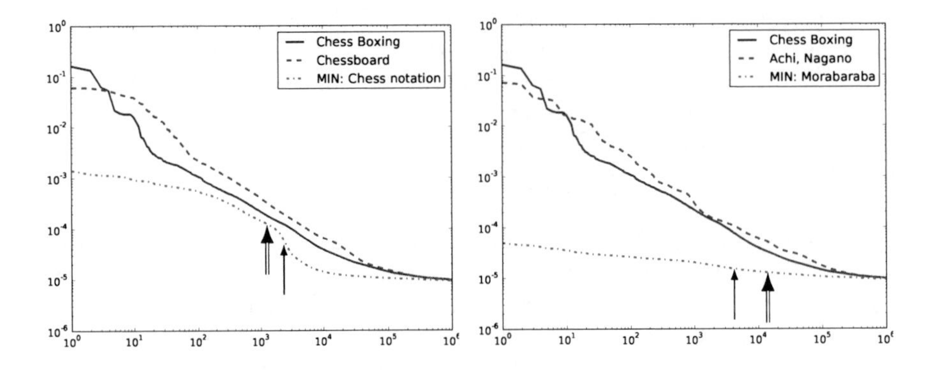

Fig. 1 Each figure shows the curves corresponding to a trial: the y axis represents the scores and the x axis represents the ranking of the nodes according to their scores. The first (resp. second) curve is the carryover opinion run from the node Chess Boxing (resp. a candidate for v, the legend shows the label of the candidate), while the third curve shows the minimum, the label of the first ranked node is in the legend. The first trial is successful, while the second is not (no plateau/decrease structure). The double arrow shows the position of the "Chess Boxing" node, while the simple arrow shows the position of the sharpest slope.

are actually mistakes of the plateau/decrease detection part of the algorithm and these groups are automatically deleted during the cleaning step.

This decomposition into 5 main groups is easily obtained by intersecting similar groups (we used a Jaccard similarity threshold of 0.7, while the other six singleton groups are automatically deleted. The labels and sizes of the 5 groups are "Enki Bilal" (35 nodes), "Uuno Turhapuro" (26 nodes), "Da Mystery of Chessboxin' " (254 nodes), "Gloria" (55 nodes) and "Queen's Gambit" (1.619 nodes). As we can see the algorithm identifies groups with very different sizes (from 26 nodes to 1.619 nodes on this example) which is a positive feature since other approaches are quite often limited to small sized communities.

Some labels are intriguing, however by checking their meaning on wikipedia on-line, all of them can be justified very easily:

- Enki Bilal is a French cartoonist. Wikipedia indicatess that "Bilal wrote [...] Froid Équateur [...] acknowledged by the inventor of chess boxing, Iepe Rubingh as the inspiration for the sport". The nodes in this group are mostly composed of its other cartoons.
- Uuno Turhapuro, is a Finnish movie. It is, as Enki Bilal, also acknowledged as the inspiration of the sport, with a scene "where the hero plays blindfold chess against one person using a hands-free telephone headset while boxing another person". The nodes in this group are mostly other cartoons characters or actors in the movies or strongly related to finnish movies.

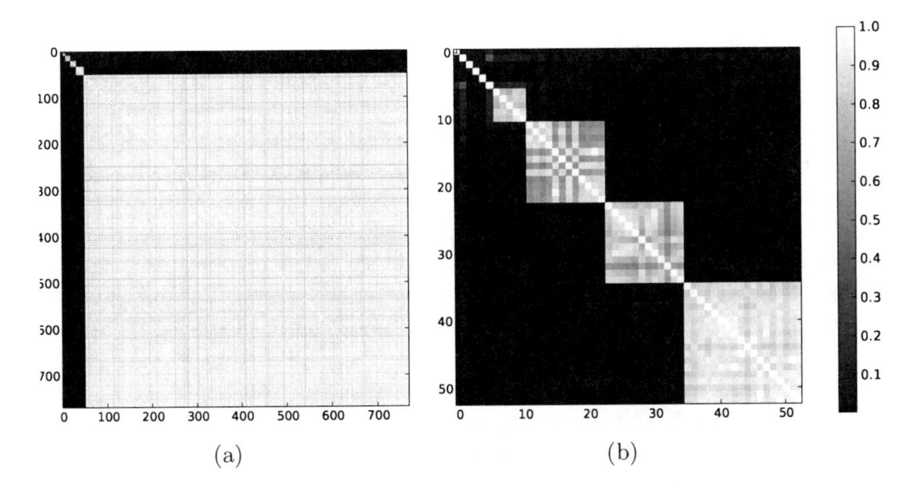

Fig. 2 Figure 2a is the rearranged Jaccard similarity matrix of these 770 communities. We see that there are 716 communities very similar to one another while not similar to the rest of the communities (the big white square). Figure 2b shows a zoom on the top left corner of the matrix.

- "Da Mystery of Chessboxin' " is a song by an American rap band: "The Wu-Tang Clan". The nodes in the community are related to the band and rap music, which is also relevant.
- "Gloria" is a page of disambiguation linking to many pages containing Gloria in their title. The current wikipedia page of "Chess Boxing" contains the sentence "On April 21, 2006, 400 spectators paid to watch two chess boxing matches in the Gloria Theatre, Cologne". However there is no hyperlink to the page "Gloria Theatre, Cologne" which is a stub. Looking at the records of wikipedia, we found that a link towards the page Gloria was added to the page "Chess Boxing" on May, the 3 2006 and then removed on January, the 31 2008. Due to the central nature of the page "Gloria" within the Gloria community, "Chess Boxing" was part of the Gloria community between these two dates, i.e., when the dataset was compiled!
- Finally, "Queen's Gambit" is a famous Chess opening and the community is composed of Chess related nodes. Even though we could have liked to label this community "Chess", "Queens' Gambit" is very specific to chess and thus characterizes this community very well.

Surprisingly, the algorithm did not find any community related to boxing. This could be a mistake due to the algorithm itself, however the wikipedia page of "Chess Boxing" explains that most chess boxers come from a chess background and learn boxing afterwards. They could thus be important within the community of Chess, but less important within the boxing community. Therefore this could explain that the "Chess Boxing" node lies within the community of Chess, but is at the limit of the boxing community.

4 Conclusion and Perspectives

We introduced an algorithm which, given a node, finds communities ego-centered on that node. Contrary to other existing algorithms our algorithm does not follow an "optimization of a quality function approach", but rather searches for irregularities in the decrease on the values of a similarity measure and leads to the detection of communities of various sizes. It also finds a practical use of the concept of multi-ego-centered communities. The algorithm is time efficient and is able to deal with very large graphs. We validated the results on a practical example using a real very large graph of wikipedia pages.

Some features of the algorithm can be improved. For instance the detection of irregularities finds only the sharpest decrease, it would be good to find all relevant irregularities, which would give multi-scale communities.

Furthermore, the algorithm is only looking for bi-centered communities, and some communities might appear only when centered on 3 or more nodes. It would be good to incorporate this feature, however it will increase the running time of the algorithm, especially because of unsuccessful trials. More advanced selection of candidates thus needs to be developed. We could for instance add the following selection feature: if a candidate is chosen for v, nodes too similar to this candidate might be neglected since they would probably lead to the same result. The speed of the algorithm is a very important feature and is central to make it practical for the study of evolving communities.

As we saw the algorithm can have some difficulties to find very small communities if there exist very large ones. This might be the reason why when applied on a globally popular node, like "Biology" or "Europe", the algorithm is only returning one very big community, while we expect to have the communities of various sub-fields of Biology or European country related topics. This is a feature of the algorithm that should be improved: relaunching the algorithm again on the induced subgraph of the nodes belonging to the big communities detected, or removing the nodes belonging to the big communities from the graph and running the algorithm again should be investigated.

Acknowledgements. This work is supported in part by the French National Research Agency contract DynGraph ANR-10-JCJC-0202.

References

[BLO08] Vincent, D.: Blondel, Jean-Loup Guillaume, Renaud Lambiotte and Etienne Lefebvre. Fast unfolding of communities in large networks. J. Stat. Mech. (2008)

[CLA05] Clauset, A.: Finding local community structure in networks. Physical Review E 72, 026132 (2005)

[CHE09] Chen, J., Zaiane, O.R., Goebel, R.: Community Identification in Social Networks. Local, Advances in Social Network Analysis and Mining (2009)

[DAN12] Danisch, M., Guillaume, J.-L., Le Grand, B.: Towards multi-ego-centered communities: a node similarity approach. Int. J. of Web Based Communities (2012)

[EVA09] Evans, T.S., Lambiotte, R.: Line Graphs, Link Partitions and Overlapping Communities. Phys.Rev.E 80, 016105 (2009), doi:10.1103/PhysRevE.80.016105

[FOR10] Fortunato, S.: Community detection in graphs. Physics Reports 486, 75–174 (2010)

[FRI11] Friggeri, A., Chelius, G., Fleury, E.: Triangles to Capture Social Cohesion. IEEE (2011)

[GIR02] Girvan, M., Newman, M.E.J.: Community structure in social and biological networks. PNAS 99(12), 7821–7826 (2002)

[GLE03] Gleiser, P., Danon, L.: Adv. Complex Syst. 6, 565 (2003)

[NEW06] Newman, M.E.J.: Finding community structure in networks using the eigenvectors of matrices. Physical Review E (2006)

[NGO12] Ngonmang, B., Tchuente, M., Viennet, E.: Local communities identification in social networks. Parallel Processing Letters 22(1) (March 2012)

[PAL05] Palla, G., Derenyi, I., Farkas, I., Vicsek, T.: Uncovering the overlapping community structure of complex networks in nature and society. Nature (2005)

[PAL08] Palla, G., Farkas, I.J., Pollner, P., Derenyi, I., Vicsek, T.: Fundamental statistical features and self-similar properties of tagged networks. New J. Phys. 10, 123026 (2008)

[ROS08] Rosvall, M., Carl, T.: Bergstrom/ Maps of information flow reveal community structure in complex networks. PNAS 105, 1118 (2008)

Application of Semidefinite Programming to Maximize the Spectral Gap Produced by Node Removal

Naoki Masuda, Tetsuya Fujie, and Kazuo Murota

Abstract. The smallest positive eigenvalue of the Laplacian of a network is called the spectral gap and characterizes various dynamics on networks. We propose mathematical programming methods to maximize the spectral gap of a given network by removing a fixed number of nodes. We formulate relaxed versions of the original problem using semidefinite programming and apply them to example networks.

Keywords: combinatorial optimization, network, synchronization, random walk, opinion formation, Laplacian, eigenvalue.

1 Introduction

An undirected and unweighted network (i.e., graph) on N nodes is equivalent to an $N \times N$ symmetric adjacency matrix $A = (A_{ij})$, where $A_{ij} = 1$ when nodes (also called vertices) i and j form a link (also called edge), and $A_{ij} = 0$ otherwise. We define the Laplacian matrix of the network by

$$L \equiv D - A, \tag{1}$$

Naoki Masuda · Kazuo Murota
Department of Mathematical Informatics, The University of Tokyo,
7-3-1 Hongo, Bunkyo, Tokyo 113-8656, Japan

Naoki Masuda
PRESTO, Japan Science and Technology Agency,
4-1-8 Honcho, Kawaguchi, Saitama 332-0012, Japan

Tetsuya Fujie
Graduate School of Business, University of Hyogo,
8-2-1 Gakuen-Nishimachi, Nishi-ku, Kobe 651-2197, Japan
e-mail: masuda@stat.t.u-tokyo.ac.jp
http://www.stat.t.u-tokyo.ac.jp/~masuda/

G. Ghoshal et al. (Eds.): *Complex Networks IV*, SCI 476, pp. 155–163.
DOI: 10.1007/978-3-642-36844-8_15

where D is the $N \times N$ diagonal matrix in which the ith diagonal element is equal to $\sum_{j=1}^{N} A_{ij}$, i.e., the degree of node i.

When the network is connected, the eigenvalues of L satisfy

$$\lambda_1 = 0 < \lambda_2 \leq \cdots \leq \lambda_N. \tag{2}$$

The eigenvalue λ_2 is called spectral gap or algebraic connectivity and characterizes various dynamics on networks including synchronizability [3, 4, 10], speed of synchronization [3], consensus dynamics [17], the speed of convergence of the Markov chain to the stationary density [8, 10], and the first-passage time of the random walk [10]. Because a large λ_2 is often considered to be desirable, e.g., for strong synchrony and high speed of convergence, maximization of λ_2 by changing networks under certain constraints is important in applications.

In the present work, we consider the problem of maximizing the spectral gap by removing a specified number, N_{del}, of nodes from a given network. We assume that an appropriate choice of N_{del} nodes keeps the network connected. A heuristic algorithm for this task in which nodes are sequentially removed is proposed in [20]. In this study, we explore a mathematical programming approach. We propose two algorithms using semidefinite programming and numerically compare their performance with that of the sequential algorithm proposed in [20].

2 Methods

We start by introducing notations. First, the binary variable x_i $(1 \leq i \leq N)$ takes a value of 0 if node i is one of the N_{del} removed nodes and 1 if node i survives the removal. Our goal is to determine x_i $(1 \leq i \leq N)$ that maximizes λ_2 under the constraint

$$\sum_{i=1}^{N} x_i = N - N_{\text{del}}. \tag{3}$$

Second, we define $\tilde{L}_{ij}$ as the $N \times N$ Laplacian matrix generated by a single link $(i, j) \in E$, where E is the set of links. In other words, the (i,i) and (j,j) elements of $\tilde{L}_{ij}$ are equal to 1, the (i,j) and (j,i) elements of $\tilde{L}_{ij}$ are equal to -1, and all the other elements of $\tilde{L}_{ij}$ are equal to 0. It should be noted that

$$L = \sum_{1 \leq i < j \leq N; (i,j) \in E} \tilde{L}_{ij}. \tag{4}$$

Third, J denotes the $N \times N$ matrix in which all the N^2 elements are equal to unity. Fourth, E_i denotes the $N \times N$ diagonal matrix in which the (i, i) element is equal to unity and all the other $N^2 - 1$ elements are equal to 0.

After the removal of $N_{\rm del}$ nodes, we do not decrease the size of the Laplacian. Instead, we remove $\tilde{L}_{ij}$ from the summation on the RHS of Eq. (4) if node i or j has been removed from the network. The Laplacian of the remaining network, if connected, has $N_{\rm del}+1$ zero eigenvalues. The corresponding zero eigenvectors are given by $\boldsymbol{u}^{(0)} \equiv (1 \cdots 1)^\top$ and $\boldsymbol{e}_i$, where $\top$ denotes the transposition, $\boldsymbol{e}_i$ is the unit column vector in which the ith element is equal to 1 and the other $N-1$ elements are equal to 0, and i is the index of one of the $N_{\rm del}$ removed nodes.

We formulate a nonlinear eigenvalue optimization problem, which we call EIGEN, as follows:

$$\text{maximize } t \quad \text{subject to Eq. (3) and}$$

$$-tI + \sum_{i<j;(i,j)\in E} x_i x_j \tilde{L}_{ij} + \alpha J + \beta \sum_{i=1}^{N} (1-x_i) E_i \succeq 0, \tag{5}$$

and $x_i \in \{0,1\}$ $(1 \le i \le N)$, where $\succeq 0$ indicates that the LHS is a positive semidefinite matrix. The positive semidefinite constraint Eq. (5) is derived from a standard prescription in semidefinite programming for optimization of an extreme eigenvalue of a matrix. Maximizing t is equivalent to maximizing the smallest eigenvalue of the matrix given by the sum of the second, third, and fourth terms on the LHS of Eq. (5).

Without the third and fourth terms on the LHS of Eq. (5), the optimal solution would be trivially equal to $t=0$ because the Laplacian of any network has 0 as the smallest eigenvalue. Because $J = \boldsymbol{u}^{(0)}\boldsymbol{u}^{(0)\top}$, the third term transports a zero eigenvalue to $\approx \alpha$. We should take a sufficiently large $\alpha > 0$ such that the zero eigenvalue is shifted to a value larger than the spectral gap of the remaining network, denoted by $\tilde{\lambda}_2$. This technique was introduced in [9] for solving the traveling salesman problem.

For each removed node i (i.e., $x_i = 0$), the matrix represented by the second term on the LHS of Eq. (5) has a zero eigenvalue associated with eigenvector $\boldsymbol{e}_i$. The fourth term shifts this zero eigenvalue to $\approx \beta$. Note that the fourth term disappears for the remaining $N - N_{\rm del}$ nodes because $x_i = 1$ for the remaining nodes. If the shifted eigenvalues are larger than $\tilde{\lambda}_2$, the solution to the problem stated above returns the $N_{\rm del}$ nodes whose removal maximizes $\tilde{\lambda}_2$.

The second term on the LHS of Eq. (5) represents a nonlinear constraint. To linearize the problem in terms of the variables, we follow a conventional prescription to introduce auxiliary variables

$$X_{ij} \equiv x_i x_j, \tag{6}$$

where $1 \le i \le j \le N$ [13, 14, 15] (also reviewed in [12]). If x_i is discrete, $x_i(1-x_i) = 0$ holds true. Therefore, we require $X_{ii} = x_i^2 = x_i$. In the following discussion, we use x_i in place of X_{ii}.

We define the $(N+1) \times (N+1)$ matrix

$$Y \equiv \begin{bmatrix} 1 & \boldsymbol{x}^\top \\ \boldsymbol{x} & X \end{bmatrix}, \tag{7}$$

where $\boldsymbol{x} \equiv (x_1 \ldots x_N)^\top$, the (i,i) element of the $N \times N$ matrix X is equal to x_i, and the (i,j) element $(i \neq j)$ of X is equal to X_{ij}. By allowing x_i and X_{ij} $(1 \leq i < j \leq N)$ to take any continuous value between 0 and 1, we define the relaxed problem named SDP1 as follows:

$$\text{maximize}\, t \quad \text{subject to Eq. (3) and}$$

$$-tI + \sum_{i<j;(i,j)\in E} X_{ij}\tilde{L}_{ij} + \alpha J + \beta \sum_{i=1}^{N} (1-x_i)E_i \succeq 0, \tag{8}$$

$$Y \succeq 0. \tag{9}$$

Note that Eq. (9) implies $0 \leq x_i \leq 1$ $(1 \leq i \leq N)$ and that SDP1 relaxes the original problem in that x_i and X_{ij} are allowed to take continuous values while Eq. (9) is imposed. The method that we propose here for approximately maximizing the spectral gap is to remove the N_{del} nodes corresponding to the N_{del} smallest values among $x_1, \ldots, x_N$ in the optimal solution of SDP1.

SDP1 involves $N(N+1)/2+1$ variables (i.e., t, x_i, and X_{ij} with $i<j$). In fact, X_{ij} for $(i,j) \notin E$ is free unless Eq. (9) is violated; it does not appear in the main positive semidefinite constraint represented by Eq. (8). Because a given network is typically sparse, this implies that there are many redundant variables in SDP1. To exploit the sparsity and thus to save time and memory space, a technique based on matrix completion might be useful [11, 16]. In this paper, however, we propose another relaxation SDP2 for this purpose.

To linearize the second term on the LHS of Eq. (5), we take advantage of four inequalities $x_i x_j \geq 0$, $x_i(1-x_j) \geq 0$, $(1-x_i)x_j \geq 0$, and $(1-x_i)(1-x_j) \geq 0$ that must be satisfied for any link $(i,j) \in E$. By defining $X_{ij} \equiv x_i x_j$, as in the case of SDP1, we obtain the following four linear constraints [18]:

$$X_{ij} \geq 0, \tag{10}$$

$$x_i - X_{ij} \geq 0, \tag{11}$$

$$x_j - X_{ij} \geq 0, \tag{12}$$

$$1 - x_i - x_j + X_{ij} \geq 0. \tag{13}$$

SDP2 is defined by replacing Eq. (9) by Eqs. (10)–(13), where only the pairs $(i,j) \in E$ are considered. Note that Eqs. (10)–(13) guarantee $0 \leq x_i \leq 1$ $(1 \leq i \leq N)$. We remove the N_{del} nodes corresponding to the N_{del} smallest values among $x_1, \ldots, x_N$ in the optimal solution of SDP2.

Numerically, SDP2 is much easier to solve than SDP1 for two reasons. First, the number of variables is smaller in SDP2 than in SDP1. In SDP2, X_{ij} is defined only on the links, whereas in SDP1 it is defined for all the pairs $1 \le i < j \le N$. In sparse networks, the number of variables is $O(N^2)$ for SDP1 and $O(N)$ for SDP2. Second, the positive semidefinite constraint, which is much more time consuming to solve than a linear constraint of a comparable size, is smaller in SDP2 than in SDP1. While SDP1 and SDP2 share the $N \times N$ positive semidefinite constraint (8), SDP1 involves an additional positive semidefinite constraint (9) of size $(N+1) \times (N+1)$.

To determine the values of α and β, we consider the matrix represented by the sum of the second, third, and fourth terms on the LHS of Eq. (5). A straightforward calculation shows that the eigenvalues of this matrix are given by the $N - N_{\text{del}} - 1$ positive eigenvalues of the Laplacian of the remaining network, $(N_{\text{del}} - 1)$-fold β, and $\beta + \left[\alpha N - \beta \pm \sqrt{(\alpha N - \beta)^2 + 4N_{\text{del}}\alpha\beta}\right]/2$. For a fixed β, we should select α to maximize $\beta + \left[\alpha N - \beta - \sqrt{(\alpha N - \beta)^2 + 4N_{\text{del}}\alpha\beta}\right]/2$, which is always smaller than eigenvalue β. We set

$$\alpha = \frac{\beta}{N} \tag{14}$$

to simplify the expression of this eigenvalue to $\beta(1 - \sqrt{N_{\text{del}}/N})$ while approximately maximizing this eigenvalue.

We have the following bounds for the optimal solution to the original problem. We denote by $\tilde{\lambda}_2^{\text{opt}}$ the optimal solution, i.e., the maximum spectral gap with N_{del} nodes removed. We denote by $\tilde{\lambda}_2^{\text{SDP}}$ the smallest positive eigenvalue of the network obtained by the proposed method; the proposed method removes the N_{del} nodes corresponding to the N_{del} smallest values of $x_1, \ldots, x_N$ in the optimal solution of SDP1 or SDP2. Obviously, $\tilde{\lambda}_2^{\text{SDP}}$ is a lower bound for $\tilde{\lambda}_2^{\text{opt}}$. On the other hand, the optimal value, $\max t$, of SDP1 or SDP2 serves as an upper bound for $\tilde{\lambda}_2^{\text{opt}}$, as long as the β satisfies $\tilde{\lambda}_2^{\text{opt}} \le \beta(1 - \sqrt{N_{\text{del}}/N})$. This follows from the facts that the optimal value of EIGEN with such a β value coincides with $\tilde{\lambda}_2^{\text{opt}}$ and both SDP1 and SDP2 are a relaxation of EIGEN. We can summarize our observation as follows: $\tilde{\lambda}_2^{\text{SDP}} \le \tilde{\lambda}_2^{\text{opt}} \le \max t$.

3 Numerical Results

In this section, we apply SDP1 and SDP2 to some synthetic and real networks. We implement SDP1 and SDP2 using the free software package SeDuMi 1.3 that runs on MATLAB 7.7.0.471 (R2008b) [1].

We compare the performance of SDP1 and SDP2 with that of the optimal sequential method, which is a heuristic method proposed in [20]. In the optimal sequential method, we numerically calculate the spectral gap for the network obtained by the removal of one node; we do this for all possible choices of a node to be removed. Subsequently, we remove the node whose removal yields the largest spectral gap. Then, for the remaining network composed of $N-1$ nodes, we determine the second node to be removed in the same way. We repeat this procedure until $N_{\rm del}$ nodes have been removed.

The first example network is the well-known karate club social network, in which a node represents a member of the club and a link represents casual interaction between two members [21]. The network has $N = 34$ nodes and 78 links. We set $\beta = 2$. The spectral gaps obtained by the different node removal methods are shown in Fig. 1(a) as a function of $N_{\rm del}$. Up to $N_{\rm del} = 5$, the optimal sequential method yields the exact solution, as do SDP1 and SDP2. For $N_{\rm del} \geq 6$, we could not obtain the exact solution by the exhaustive search because of the combinatorial explosion. For $7 \leq N_{\rm del} \leq 16$, SDP1 and SDP2 perform worse than the optimal sequential method. However, for $N_{\rm del} \geq 17$, both SDP1 and SDP2 outperform the optimal sequential method. SDP1 and SDP2 found efficient combinations of removed nodes that the optimal sequential method could not find.

Second, we test the three methods against the largest connected component of the undirected and unweighted version of a macaque cortical network [19]. The network has $N = 71$ nodes and 438 links. We set $\beta = 2$. The spectral gaps obtained by the different methods are shown in Fig. 1(b). Up to $N_{\rm del} = 4$, the optimal sequential method yields the exact solution, as do SDP1 and SDP2. For $N_{\rm del} \geq 5$, we could not obtain the exact solution because of the combinatorial explosion. For $N_{\rm del} \geq 5$, SDP1 and SDP2 perform worse than the optimal sequential method. Consistent with the poor performance of SDP1 and SDP2, the final values of x_i $(1 \leq i \leq N)$ are not bimodally distributed around 0 and 1 as SDP1 and SDP2 implicitly suppose. The distribution is rather unimodal except for the first three values of x_i that are close to 0. The ten values of x_i when $N_{\rm del} = 5$, in ascending order, are as follows: $x_{33} = 0.1086$, $x_{62} = 0.1531$, $x_{53} = 0.1589$, $x_1 = 0.4813$, $x_2 = 0.5246$, $x_8 = 0.5591$, $x_7 = 0.6449$, $x_{24} = 0.7866$, $x_{51} = 0.8749$, and $x_{63} = 0.8931$ in SDP1, and $x_{53} = 0.000$, $x_{33} = 0.145$, $x_{62} = 0.177$, $x_2 = 0.585$, $x_1 = 0.588$, $x_8 = 0.610$, $x_7 = 0.668$, $x_{24} = 0.708$, $x_5 = 0.738$, and $x_4 = 0.937$ in SDP2.

The third network is a network with $N = 150$ nodes generated by the Barabási–Albert scale-free network model [5]. The growth of the network starts with a connected pair of nodes, and each incoming node is assumed to have two links. The generated network has 297 links. We set $\beta = 2$. For this and the next networks, SDP1 cannot be applied because N is too large. Therefore, we only compare the performance of SDP2 against the optimal sequential method. The results shown in Fig. 1(c) indicate that SDP2 outperforms the optimal sequential method when $N_{\rm del} \geq 7$.

The fourth network is the largest connected component of the *C. elegans* neural network [2,7]. Two nodes are regarded as being connected when they are connected by a chemical synapse or gap junction. We ignore the direction and weight of links. The network has $N = 279$ nodes and 2287 links. We set

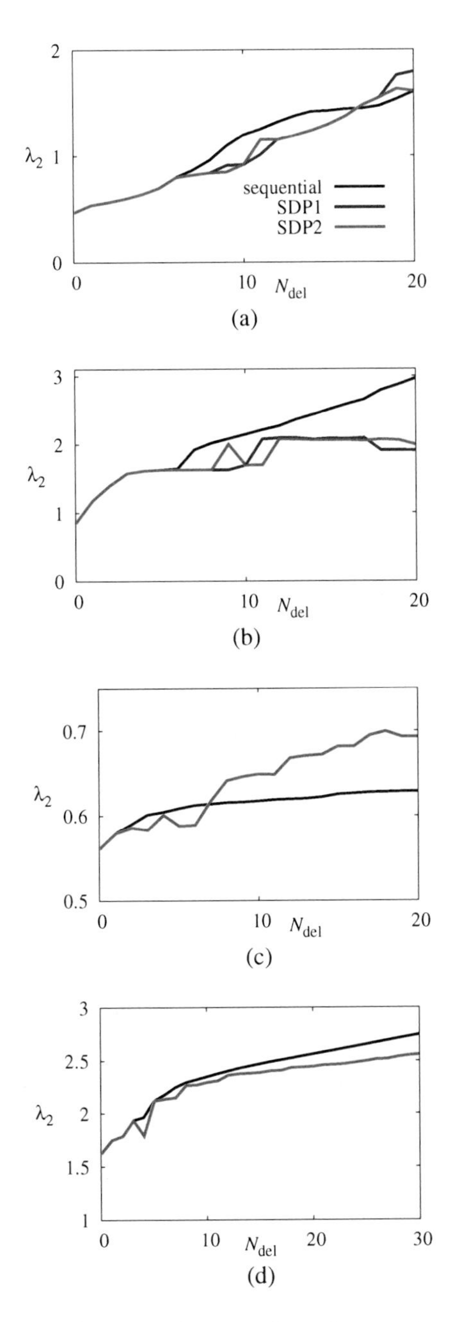

Fig. 1 Spectral gap as a function of the number of removed nodes for four networks. (a) Karate club social network with $N = 34$ nodes. (b) Macaque cortical network with $N = 71$ nodes. (c) Barabási–Albert scale-free network with $N = 150$ nodes. (d) *C. elegans* neural network with $N = 279$ nodes.

$\beta = 2.5$. The results for SDP2 and the optimal sequential method are shown in Fig. 1(d). Although the spectral gap gradually increases with N_{del} for SDP2, SDP2 performs poorly as compared to the optimal sequential method for this example.

4 Discussion

We proposed a method to maximize the spectral gap using semidefinite programming. The two proposed algorithms have a firmer mathematical foundation as compared to the heuristic numerical method (i.e., optimal sequential method). The proposed algorithms performed better than the heuristic method for two networks especially for large N_{del} and worse for the other two networks. For the former two networks, we could find the solutions in the situations in which the exhasutive search is computationally formidable. Up to our numerical efforts, our algorithms seem to be efficient for sparse networks.

We should be careful about the choice of β. If β is too large, SDP1 and SDP2 would result in $x_i \approx N_{\mathrm{del}}/N$ $(1 \le i \le N)$. This is because setting $x_i = N_{\mathrm{del}}/N$ $(1 \le i \le N)$ makes the fourth term on the LHS of Eq. (5) equal to $\beta \frac{N-N_{\mathrm{del}}}{N} I$, which increases all the eigenvalues, including the spectral gap of the remaining network, by $\beta \frac{N-N_{\mathrm{del}}}{N}$. In contrast, if β is smaller than $\tilde{\lambda}_2$, SDP1 and SDP2 would maximize a false eigenvalue originating from the fourth term on the LHS of Eq. (5).

To enhance the performance of SDP1 and SDP2, it may be useful to abandon the convexity of the problem. For example, we could try replacing $(1-x_i)$ in the fourth term by $(1-x_i)^p$ and gradually increase p from unity. When $p > 1$, the problem is no longer convex. Accordingly, the existence of the unique solution and the convergence of a proposed algorithm are not guaranteed. Nevertheless, we may be able to track the optimal solution $\boldsymbol{x}$ by the Newton method while we gradually increase p (see p.5 and p.63 in [6]). An alternative extension is to add $-p\sum_{i=1}^{N} x_i(1-x_i)$ to the objective function to be maximized (i.e., t). When $p > 0$, the convexity is violated. However, we may be able to adopt a procedure similar to the method explained above, i.e., start with $p = 0$ and gradually increase p to track the solution by the Newton method.

Acknowledgements. Naoki Masuda acknowledges the financial support of the Grants-in-Aid for Scientific Research (no. 23681033) from MEXT, Japan. This research is also partially supported by the Aihara Project, the FIRST program from JSPS and by Global COE Program "The research and training center for new development in mathematics" from MEXT.

References

1. http://sedumi.ie.lehigh.edu
2. http://www.wormatlas.org
3. Almendral, J.A., Díaz-Guilera, A.: Dynamical and spectral properties of complex networks. New J. Phys. 9, 187 (2007)
4. Arenas, A., Díaz-Guilera, A., Kurths, J., Moreno, Y., Zhou, C.: Synchronization in complex networks. Phys. Rep. 469, 93–153 (2008)
5. Barabási, A.L., Albert, R.: Emergence of scaling in random networks. Science 286, 509–512 (1999)
6. Bendsøe, M.P., Sigmund, O.: Topology Optimization. Springer (2003)
7. Chen, B.L., Hall, D.H., Chklovskii, D.B.: Wiring optimization can relate neuronal structure and function. Proc. Natl. Acad. Sci. USA 103, 4723–4728 (2006)
8. Cvetković, D., Rowlinson, P., Simić, S.: An Introduction to the Theory of Graph Spectra. CMU (2010)
9. Cvetkovic, D., Cangalovic, M., Kovacevic-Vujcic, V.: Semidefinite Programming Methods for the Symmetric Traveling Salesman Problem. In: Cornuéjols, G., Burkard, R.E., Woeginger, G.J. (eds.) IPCO 1999. LNCS, vol. 1610, pp. 126–136. Springer, Heidelberg (1999)
10. Donetti, L., Neri, F., Munoz, M.A.: Optimal network topologies: expanders, cages, Ramanujan graphs, entangled networks and all that. J. Stat. Mech., P08007 (2006)
11. Fukuda, M., Kojima, M., Murota, K., Nakata, K.: Exploiting sparsity in semidefinite programming via matrix completion I: general framework. SIAM J. Optim. 11, 647–674 (2000)
12. Goemans, M.X.: Semidefinite programming in combinatorial optimization. Math. Programming 79, 143–161 (1997)
13. Grötschel, M., Lovász, L., Schrijver, A.: Relaxations of vertex packing. J. Comb. Theory B 40, 330–343 (1986)
14. Lovász, L.: On the Shannon capacity of a graph. IEEE Trans. on Info. Th. 25, 1–7 (1979)
15. Lovász, L., Schrijver, A.: Cones of matrices and set-functions and 0–1 optimization. Siam J. Optimiz. 1, 166–190 (1991)
16. Nakata, K., Fujisawa, K., Fukuda, M., Kojima, M., Murota, K.: Exploiting sparsity in semidefinite programming via matrix completion II: implementation and numerical results. Math. Program. Ser. B 95, 305–327 (2003)
17. Olfati-Saber, R., Fax, J., Murray, R.: Consensus and cooperation in networked multi-agent systems. Proceedings of the IEEE 95, 215–233 (2007)
18. Padberg, M.: The Boolean quadric polytope—some characteristics, facets and relatives. Math. Programming 45(1), 139–172 (1989)
19. Sporns, O., Zwi, J.D.: The small world of the cerebral cortex. Neuroinformatics 4, 145–162 (2004)
20. Watanabe, T., Masuda, N.: Enhancing the spectral gap of networks by node removal. Phys. Rev. E 82, 46102 (2010)
21. Zachary, W.W.: An information flow model for conflict and fission in small groups. J. Anthropological Res. 33, 452–473 (1977)

Singularities in Ternary Mixtures of k-core Percolation

Davide Cellai and James P. Gleeson

Abstract. Heterogeneous k-core percolation is an extension of a percolation model which has interesting applications to the resilience of networks under random damage. In this model, the notion of node robustness is local, instead of global as in uniform k-core percolation. One of the advantages of k-core percolation models is the validity of an analytical mathematical framework for a large class of network topologies. We study ternary mixtures of node types in random networks and show the presence of a new type of critical phenomenon. This scenario may have useful applications in the stability of large scale infrastructures and the description of glass-forming systems.

1 Introduction

Percolation, with its many modifications and extensions, is a problem with a venerable past and many applications in the most diverse disciplines [18]. With the rapid development of network science, several new problems and models have been introduced. Among the different directions, k-core percolation constitutes a development which, in spite of its somewhat simple definition, is able to encapsulate a number of interesting problems which can often be approached with a robust mathematical formalism. Given a network, a k-core is defined as the subnetwork where each node has at least k neighbours in the same subnetwork. A k-core can equivalently be defined as the subnetwork remaining after a culling process consisting in recursively removing all the nodes with degree lower than k. k-core percolation has applications in many different disciplines including jamming [15], neural networks [8], granular gases [1], evolution [14], social sciences [13] and the metal-insulator

Davide Cellai · James P. Gleeson
MACSI, Department of Mathematics and Statistics, University of Limerick, Ireland
e-mail: {davide.cellai,james.gleeson}@ul.ie

G. Ghoshal et al. (Eds.): *Complex Networks IV*, SCI 476, pp. 165–172.
DOI: 10.1007/978-3-642-36844-8_16

transition [4]. The application of k-core percolation we are interested in here, however, refers to the stability of a network under random damage. A fraction $(1-p)$ of nodes is removed (together with the adjacent edges) and the size M_k of the largest k-core cluster is studied as a function of p. The k-core strength M_k can vanish either continuously or discontinuously, a scenario that can be described in terms of phase transitions (second and first order, respectively) and can be important in the stability of large scale infrastructures [9, 10]. A recent extension of k-core percolation, named heterogeneous k-core (HKC) percolation considers the threshold k_i as a local property, and allows therefore the mixing of different critical phenomena. An analytical formalism has been introduced to approach this model on networks and binary mixtures of thresholds k have been investigated [2, 5, 6]. On a wide class of networks, this model displays phase diagrams which are topologically equivalent to the ones calculated for a recently introduced spin model of glass-forming systems [17, 16]. This model can be seen as a heterogeneous development of the Fredrickson-Andersen (FA) model of facilitated spins, where the facilitation consists of a local constraint in the number of spins down in order for the considered spin to be able to flip [11, 17]. It has been shown that this model reproduces characteristic signatures of glass-forming systems called *glass-transition singularities*, which correspond to distinctive critical phenomena for the appropriate choice of the parameters [12, 16]. One of the glass-transition singularities which has not yet been explored is the so-called A_4 singularity, characterised by the coalescence of two critical points. As the HKC model appears to reproduce all the relevant critical phenomena for binary mixtures, it is interesting to look for this singularity as well. The purpose of this paper, therefore, is to investigate, for the first time, ternary mixtures in HKC percolation and calculate the critical point corresponding to an A_4 singularity.

In Section 2 we present the analytical formalism used here (from [2]), in Section 3 we sketch the behaviour of the model for binary mixtures, in Section 4 we show a case of ternary mixture with an A_4 singularity and in Section 5 we give the conclusions.

2 Formalism

Let k_i be the k-core threshold of the node i. A fraction $(1-p)$ of nodes are randomly removed: the problem consists in calculating the size of the HKC (if it exists). We consider the configuration model of random networks, defined as the maximally random network with a given degree distribution $P(q)$. The configuration model has the property of being locally tree-like, i. e. the number of finite loops vanishes for infinite networks. This property allows to consider, in an infinite network, the HKC equivalent to the (k_i-1)-ary subtree, defined as the tree in which, as we traverse it, each encountered vertex has at least k_i-1 child edges. Then we can write a self-consistent equation for Z, the probability that a randomly chosen node is the root of a (k_i-1)-ary subtree [2]:

$$
\begin{aligned}
Z = pr \sum_{q=k_a}^{\infty} \frac{qP(q)}{\langle q \rangle} \sum_{l=k_a-1}^{q-1} \binom{q-1}{l} Z^l (1-Z)^{q-1-l} + \\
+ps \sum_{q=k_b}^{\infty} \frac{qP(q)}{\langle q \rangle} \sum_{l=k_b-1}^{q-1} \binom{q-1}{l} Z^l (1-Z)^{q-1-l} + \\
+p(1-r-s) \sum_{q=k_c}^{\infty} \frac{qP(q)}{\langle q \rangle} \sum_{l=k_c-1}^{q-1} \binom{q-1}{l} Z^l (1-Z)^{q-1-l}.
\end{aligned} \tag{1}
$$

where the three thresholds $\mathbf{k} = (k_a, k_b, k_c)$ are randomly assigned to nodes with probability r, s and $(1-r-s)$, respectively. In this paper we always assume $k_i \geq 2$, so we do not need to consider the case where there may be finite clusters in the HKC. Due to the absence of finite loops, no finite HKC clusters can exist if $k_i \geq 2$ for all nodes i [10, 2].

We can then write the probability M_{abc} that a randomly chosen node is in the HKC, for a mixture of three types of nodes:

$$
\begin{aligned}
M_{abc}(p) = pr \sum_{q=k_a}^{\infty} P(q) \sum_{l=k_a}^{q} \binom{q}{l} Z^l (1-Z)^{q-l} + \\
ps \sum_{q=k_b}^{\infty} P(q) \sum_{l=k_b}^{q} \binom{q}{l} Z^l (1-Z)^{q-l} + \\
p(1-r-s) \sum_{q=k_c}^{\infty} P(q) \sum_{l=k_c}^{q} \binom{q}{l} Z^l (1-Z)^{q-l}.
\end{aligned} \tag{2}
$$

3 Binary Mixtures in Heterogeneous k-core Percolation

In homogeneous k-core percolation it is known that, for networks with a fast decreasing degree distribution, namely $P(q) < 1/q^{\gamma}$ with $\gamma > 3$ for $q \to \infty$, a randomly damaged network collapses continuously for $k \leq 2$ and discontinuously for $k \geq 3$ [7, 3, 10]. In recent papers, it has also been shown that in the case of binary mixtures HKC percolation is characterized by a few different critical phenomena. For example, in mixtures where the two values of k are associated to continuous and discontinuous transitions, respectively, a critical point (also called A_3 singularity or cusp singularity) is usually observed [2, 6]. That is the case, for instance, of the mixtures $\mathbf{k} = (1,3)$ or $\mathbf{k} = (2,4)$, where the line of first order transitions ends in a critical point and a line of second order transition intersects the former. The case $\mathbf{k} = (2,3)$, though, is quite peculiar as the critical line exactly matches the line of first order transitions giving rise to a tricritical point (Fig. 1) [5].

When both values of k are characterised by a first order transition, instead, the phase diagram displays a critical point only when the two values of k are different enough [6]. Fig. 2, for instance, shows that the mixture $\mathbf{k} = (3,8)$ has a critical point whereas $\mathbf{k} = (3,4)$ does not have one.

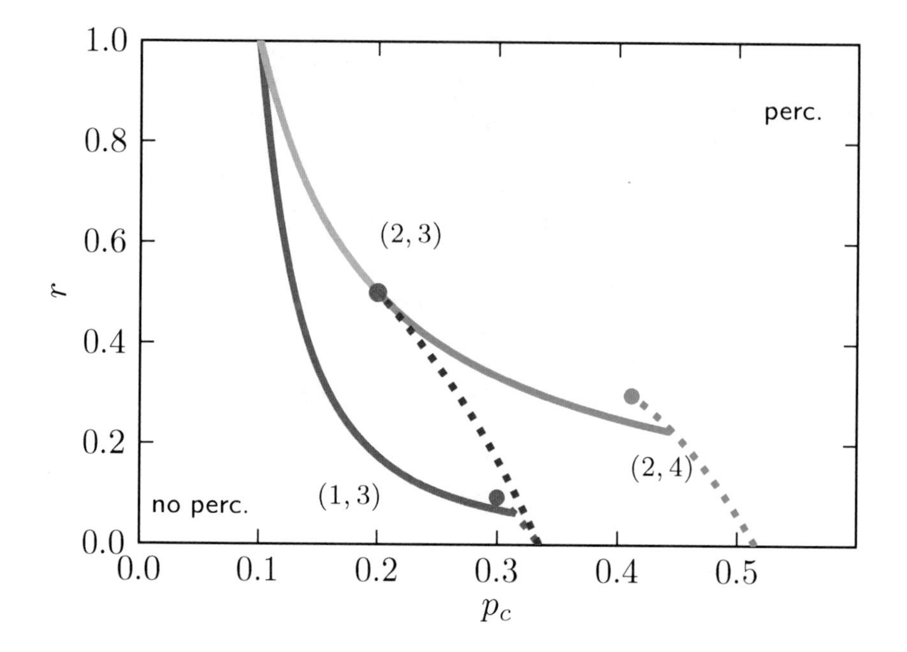

Fig. 1 Comparison of the phase diagrams of the cases $\mathbf{k} = (1,3)$, $\mathbf{k} = (2,3)$ and $\mathbf{k} = (2,4)$ for Erdős-Rényi graphs with $z_1 = 10$. Continuous lines represent second order phase transitions, whereas dashed lines represent first order phase transitions. The red dot represents a tricritical point; the other dots are critical points. In terms of the notation given in Section 2, a binary mixture corresponds for example to $s = 1 - r$.

4 The A_4 Singularity

From the analogy between HKC percolation and facilitated spin models, it is interesting to investigate whether the singularities observed in models of the glass-forming systems are also present in HKC percolation. In particular, the fact that all the critical phenomena observed so far in HKC percolation are in the same class of universality as the ones observed in the FA model with heterogeneous facilitation is quite remarkable [16, 6]. The MCT of glass-forming systems also predicts an "A_4 singularity", meaning the coalescence of a critical point into a line of first order phase transitions [12]. The A_4 singularity is also named a swallow-tail bifurcation. As this scenario is associated with a three-parameter theory, it is reasonable to expect that the corresponding singularity in our HKC model can be observed in a three-component mixture. Moreover, we have seen in Section 3 that binary mixtures with $k_i \geq 3$ are either characterised by a single line of first order transitions or two lines of first order transitions with a single critical point. It is therefore natural to consider a ternary mixture which interpolates between the two regimes. We now show that an A_4 singularity is indeed present in the ternary mixture $\mathbf{k} = (3,5,8)$ on Erdős-Rényi graphs.

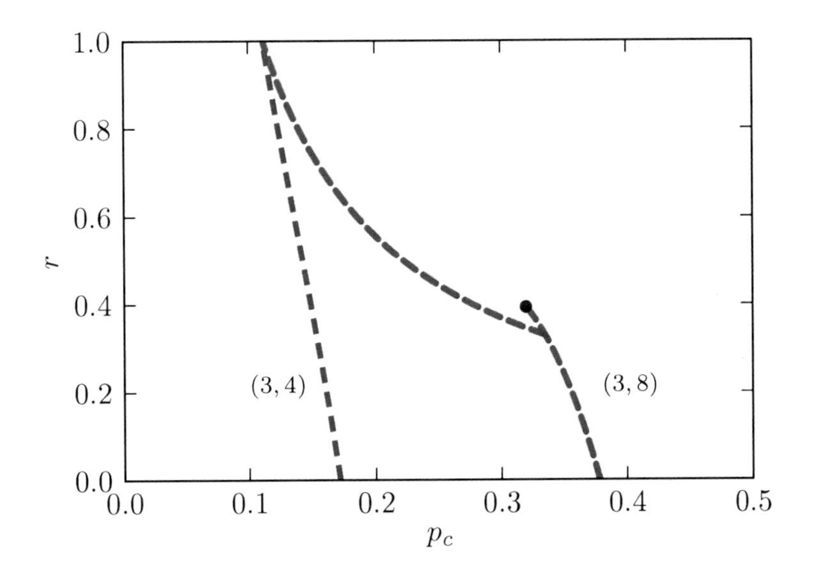

Fig. 2 Phase diagram of the cases $\mathbf{k} = (3,8)$ and $\mathbf{k} = (3,4)$ for the Erdős-Rényi graph ($z_1 = 30$). Symbols are as in Fig. 1. While for $\mathbf{k} = (3,4)$ there is only one line, in the case $\mathbf{k} = (3,8)$ there are two lines of discontinuous transitions and a critical point.

Equation (1) can be re-written as

$$f_{358}(Z) = \frac{1}{p}, \tag{3}$$

where

$$f_{358}(Z) = \frac{1}{Z}\left\{1 - e^{-z_1 Z}\left[1 + rz_1 Z + s\sum_{n=1}^{3}\frac{(z_1 Z)^n}{n!} + (1-r-s)\sum_{n=1}^{6}\frac{(z_1 Z)^n}{n!}\right]\right\}. \tag{4}$$

At every fixed fraction of remaining nodes p, depending on the values of the parameters r and s, f_{358} has one or two maxima in Z. If the second maximum is higher than the first one, there is only one first order transition between a percolating and a non-percolating HKC. If the first maximum is higher than the second one, there are two first order transitions: one between a high-k and a low-k phase and another towards a collapsing HKC. Typically, the second maximum disappears at a critical point, which can be found by imposing the condition:

$$f'_{358}(Z) = f''_{358}(Z) = 0. \tag{5}$$

This defines a locus in the (r,s) plane corresponding to a line of critical points. At high values of s, however, the critical point disappears and only a single line of first order transitions survives. This is due to the fact that the 5-nodes interpolate between the other two values of k and there is no critical point in either the mixture

$\mathbf{k} = (3,5)$ or $\mathbf{k} = (5,8)$. The condition of an A_4 singularity corresponds to the onset of this behaviour and is defined by the condition:

$$f'_{358}(Z) = f''_{358}(Z) = f'''_{358}(Z) = 0, \tag{6}$$

which yields

$$r_* = 0.3140687639806 \qquad s_* = 0.1831697392197 \tag{7}$$

which is the position of the A_4 singularity on the phase diagram. Fig. 3 shows the phase diagram in the plane (p,r) for a few values of s.

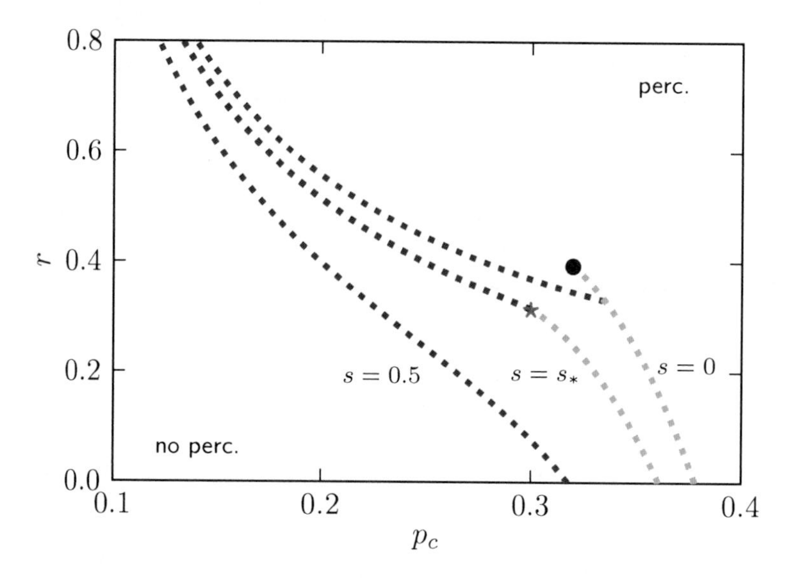

Fig. 3 Phase diagram of the mixture $\mathbf{k} = (3,5,8)$ for Erdős-Rényi graphs ($z_1 = 30$). r and s are the fractions of nodes of type 3 and 5, respectively, and p_c is the fraction of undamaged nodes at a phase transition. Lines of discontinuous transitions are plotted in the plane (p,r) at three fixed values of s. In the absence of type 5 ($s = 0$), there are two lines of first order transitions. One of them ends in a critical point (black dot). At the critical fraction s_*, the two lines touch at a A_4 singularity (red star). For $s > s_*$, there is only a single line of discontinuous transitions.

The critical exponent β is defined by the vanishing of the order parameter in approaching a phase transition: $M_{358}(p) - M_{358}(p_c) \sim (p - p_c)^\beta$. As in binary mixtures, the critical exponent at the discontinuous phase transition is $\beta = 1/2$, whereas it becomes $\beta = 1/3$ at the critical point. In the vicinity of the A_4 singularity, we have $(p - p_*) \sim f_{358}(Z_*) - f_{358}(Z) \sim (Z - Z_*)^4$, from which it follows that $M_{358}(p) - M_{358}(p_*) \sim (p - p_*)^{1/4}$, as M_{358} is linear in p and Z in the vicinity of the transition. Thus, the values of the critical exponent β can be summarized as follows:

$$\beta = \begin{cases} 1/2 \text{ hybrid transition } & r \neq r_c, s \neq s_* \\ 1/3 \text{ critical point } & r = r_c, s \neq s_* \\ 1/4 \ A_4 \text{ singularity } & r = r_*, s = s_* \end{cases} \tag{8}$$

The unique value of $\beta = 1/4$ identifies the critical point as a A_4 singularity. This phenomenon has an interesting interpretation. In a networks formed by types of nodes of very different fragilities (as in $k_a = 3$, $k_c = 8$), the collapse due to random damage can be catastrophic, with the presence of multiple discontinuous transitions. The introduction of nodes of intermediate fragility ($k_b = 5$) can weaken the critical point up to disappearance, making a significant region of the phase diagram stable (Fig. 3). This purely mathematical remark could be useful in principle in planning the stability of large scale infrastructures. Another interesting application is in modelling the glass transition. The presence of A_4 singularities is predicted by the mode-coupling theory of glass-forming systems [12]. Due to the strong similarities between HKC percolation and facilitated spin models, it is reasonable to expect that this singularity should also be present in suitable ternary mixtures of facilitation parameters in such spin models [17, 16].

5 Conclusions

In this paper we have applied the formalism of heterogeneous k-core percolation to ternary mixtures of thresholds k. In particular, we have calculated the phase diagram of a ternary mixture for Erdős-Rényi graphs. This mixture displays a characteristic A_4 singularity, i. e. a critical phenomenon characterised by the merging of a critical point with a distinct line of discontinuous transitions. A peculiarity of this point is the change in the critical exponent β of the order parameter, which uniquely assumes the value $1/4$. The characteristics of the studied phase diagrams may give useful information in designing large scale infrastructures to be resilient to random damage. This singularity is also predicted by models of the glass transition [16] and our model appears to be in the same universality class of kinetic spin models with heterogeneous facilitation.

Acknowledgements. We acknowledge useful discussions with Mauro Sellitto. This work has been funded by Science Foundation Ireland, grants: 11/PI/1026 and 06/MI/005.

References

1. Alvarez-Hamelin, J.I., Puglisi, A.: Dynamical collision network in granular gases. Phys. Rev. E 75, 51302 (2007)
2. Baxter, G.J., Dorogovtsev, S.N., Goltsev, A.V., Mendes, J.F.F.: Heterogeneous k-core versus bootstrap percolation on complex networks. Phys. Rev. E 83(5), 051134 (2011)
3. Branco, N.S.: Probabilistic bootstrap percolation. J. Stat. Phys. 70, 1035 (1993)

4. Cao, L., Schwarz, J.M.: Quantum *k*-core conduction on the bethe lattice. Phys. Rev. B 82, 104211 (2010), `http://link.aps.org/doi/10.1103/PhysRevB.82.104211`
5. Cellai, D., Lawlor, A., Dawson, K.A., Gleeson, J.P.: Tricritical Point in Heterogeneous k-Core Percolation. Physical Review Letters 107, 175703 (2011), `http://dx.doi.org/10.1103/PhysRevLett.107.175703`
6. Cellai, D., Lawlor, A., Dawson, K.A., Gleeson, J.P.: Critical phenomena in heterogeneous k-core percolation (September 2012), `http://arxiv.org/abs/1209.2928`
7. Chalupa, J., Leath, P.L., Reich, G.R.: Bootstrap percolation on a bethe lattice. J. Phys. C 12, L31 (1979)
8. Chatterjee, N., Sinha, S.: Understanding the mind of a worm: hierarchical network structure underlying nervous system function in c. elegans. Prog. Brain Res. 168, 145–153 (2007)
9. Cohen, R., Erez, K., Ben Avraham, D., Havlin, S.: Resilience of the internet to random breakdowns. Phys. Rev. Lett. 85, 4626 (2000)
10. Dorogovtsev, S.N., Goltsev, A.V., Mendes, J.F.F.: *k*-core organization of complex networks. Phys. Rev. Lett. 96(4), 40601 (2006)
11. Fredrickson, G.H., Andersen, H.C.: Kinetic ising model of the glass transition. Physical Review Letters 53(13), 1244–1247 (1984), `http://dx.doi.org/10.1103/PhysRevLett.53.1244`
12. Götze, W.: Complex Dynamics of Glass-Forming Liquids: A Mode-Coupling Theory. International Series of Monographs on Physics. Oxford University Press, USA (2009), `http://www.amazon.com/exec/obidos/redirect?tag=citeulike07-20&path=ASIN/B007PM7EKU`
13. Kitsak, M., Gallos, L.K., Havlin, S., Liljeros, F., Muchnik, L., Stanley, H.E., Makse, H.A.: Identification of influential spreaders in complex networks. Nat. Phys. 6(11), 888–893 (2010)
14. Klimek, P., Thurner, S., Hanel, R.: Pruning the tree of life: k-core percolation as selection mechanism. J. Theor. Biol. 256(1), 142–146 (2009)
15. Schwarz, J.M., Liu, A.J., Chayes, L.Q.: The onset of jamming as the sudden emergence of an infinite k-core cluster. Europhys. Lett. 73, 560 (2006)
16. Sellitto, M.: Physical Review E 86, 030502(R) (2012)
17. Sellitto, M., De Martino, D., Caccioli, F., Arenzon, J.J.: Dynamic facilitation picture of a higher-order glass singularity. Phys. Rev. Lett. 105, 265704 (2010)
18. Stauffer, D., Aharony, A.: Introduction To Percolation Theory, 2nd edn. CRC Press (July 1994), `http://www.worldcat.org/isbn/0748402535`

Assessing Particle Swarm Optimizers Using Network Science Metrics

Marcos A.C. Oliveira-Júnior, Carmelo J.A. Bastos-Filho, and Ronaldo Menezes

Abstract. Particle Swarm Optimizers (PSOs) have been widely used for optimization problems, but the scientific community still does not have sophisticated mechanisms to analyze the behavior of the swarm during the optimization process. We propose in this paper to use some metrics described in network sciences, specifically the R-value, the number of zero eigenvalues of the Laplacian Matrix, and the Spectral Density, in order to assess the behavior of the particles during the search and diagnose stagnation processes. Assessor methods can be very useful for designing novel PSOs or when one needs to evaluate the performance of a PSO variation applied to a specific problem. In order to apply these metrics, we observed that it is not possible to analyze the dynamics of the swarm by using the communication topology because it does not change. Therefore, we propose in this paper the definition of the influence graph of the swarm. We used this novel concept to assess the dynamics of the swarm. We tested our proposed methodology in three different PSOs in a well-known multimodal benchmark function. We observed that one can retrieve interesting information from the swarm by using this methodology.

1 Introduction

Computational Swarm Intelligence (SI) is a set of bio-inspired algorithms based on populations of simple reactive agents. They interact locally among themselves in other to generate global patterns that can be used to solve complex tasks [6]. Among the most famous SI algorithms, we can cite: Particle Swarm Optimization

Marcos A.C. Oliveira-Júnior · Carmelo J.A. Bastos-Filho
University of Pernambuco, Brazil
e-mail: carmelofilho@ieee.org

Ronaldo Menezes
Florida Institute of Technology, USA
e-mail: rmenzes@cs.fit.edu

G. Ghoshal et al. (Eds.): *Complex Networks IV*, SCI 476, pp. 173–184.
DOI: 10.1007/978-3-642-36844-8_17

(PSO) [9], Ant Colony Optimization (ACO) [5], Artificial Bee Colony (ABC) [8] and Fish School Search (FSS) [2].

PSO has been widely used to solve optimization problems in hyper-dimensional search spaces with continuous variables. PSO was first proposed by Kennedy and Eberhart in 1995 [9], inspired by the social behavior of flocks of birds aiming to find food. In PSO, each particle in the swarm represents a candidate solution for the optimization problem. During the algorithm execution, each particle adjusts its position based on the current position, the current velocity, the best position achieved by itself during the search process so far and the best position obtained by the best particle in its neighborhood. This neighborhood is defined by the swarm communication topology, which defines which particles can exchange information among each other. The topology influences on the convergence velocity and on the quality of the solution obtained by the algorithm [3, 10]. Less connected topologies slow down the information flow, since the information is transmitted indirectly through intermediary particles [10]. Conversely, highly connected topologies decrease the average distance between any pair of individuals. As a consequence, there is a tendency for the whole swarm to move quickly toward the first local optimum found by any particle, when the average distance between particles is too short. In order to overcome this trade-off, some dynamic self-adjustable topologies were proposed aiming to manage the information flow during the execution of the PSO [13][12]. Furthermore, Oliveira-Júnior *et al.* [12] proposed one dynamical topology based on the preferential attachment mechanism of scale-free networks [1].

Nevertheless, the analyses on the influence of the communication topology in the algorithm performance generally are performed by using tools that can not provide comprehensive information about the swarm behavior. In general, researchers use metrics that do not assess the flow of information within the swarm; they just evaluate simple metrics, such as the average distance between particles, and the evolution of the fitness of the particles along the iterations. In fact, as we look further into the literature we conclude that there are no tools to appropriately assess the communication processes within swarm.

Network Science is the study of the theoretical foundations of network structure, its dynamic behavior, and the application of networks to many subfields [11]. There are some networks that present some specific characteristics, such as Scale-Free [1] and Small-World Networks [15]. In order to classify networks based on their structural features, many metrics have been developed [7]. In this paper, we propose to use these metrics to analyze the communication behavior of the particles during the search and diagnose stagnation processes. We also present the concept of *influence graph* of the swarm and we use this concept to assess the communication among the particles.

The paper is organized as follows: we briefly review the Particle Swarm Optimization and some Network Science metrics in Section 2. In Section 3, the simulation setup and results are presented. Finally, we provide our conclusions and suggest some future works in Section 4.

2 Background

2.1 Particle Swarm Optimization

Particle Swarm Optimization (PSO) is a stochastic, bio-inspired, population-based global optimization technique [9]. In PSO, each particle i has a position at time t within the search space $\mathbf{x_i}(t)$ and each position represents a possible solution for a d-dimensional optimization problem.

The particles "fly" through the search space of the problem seeking best solutions. Each particle updates its position according to the current velocity $\mathbf{v}_i(t)$, the best position found by itself $\mathbf{P}_{best_i}(t)$ and the best position found by the neighborhood of the particle i during the search so far $\mathbf{N}_{best_i}(t)$.

The velocity and the position of every particle are updated iteratively by applying the following update equations:

$$\mathbf{v_i}(t+1) = \mathbf{v_i}(t) + r_1 c_1 [\mathbf{P}_{best_i}(t) - \mathbf{x_i}(t)] + r_2 c_2 [\mathbf{N}_{best_i}(t) - \mathbf{x_i}(t)], \quad (1)$$

$$\mathbf{x_i}(t+1) = \mathbf{x_i}(t) + \mathbf{v_i}(t+1), \quad (2)$$

in which r_1 and r_2 are random numbers generated by an uniform probability density function in the interval [0,1] at each iteration for each particle for every dimension. The learning factors c_1 and c_2 are the cognitive and the social acceleration constants. They are non-negative constants and weight the contribution of the cognitive and social components, *i.e.* the second and the third terms of Equation 1.

Clerc [4] observed that the original PSO can operate in unstable states if the parameters of Equation 1 are not selected properly and determined a relation based on the constriction factor (χ) that avoids the explosion state. χ is defined according to the following equation:

$$\chi = \frac{2}{|2 - \varphi - \sqrt{\varphi^2 - 4\varphi}|}, \ \varphi = c_1 + c_2. \quad (3)$$

The mechanism to update the velocity proposed by Clerc is presented in Equation 4.

$$\mathbf{v_i}(t+1) = \chi \cdot \{\mathbf{v_i}(t) + r_1 c_1 [\mathbf{P}_{best_i}(t) - \mathbf{x_i}(t)] + r_2 c_2 [\mathbf{N}_{best_i}(t) - \mathbf{x_i}(t)]\}. \quad (4)$$

The constriction factor was designed to adjust the influence of the previous particle velocities on the optimization process. It also helps to switch the search mode of the swarm from exploration to exploitation during the search process.

2.1.1 Particle Swarm Optimization Topologies

The way the information flows through the particles is determined by the communication topology used by the swarm. The topology defines the neighborhood of each particle, *i.e.* the subset of particles which the particle is able to communicate with.

There are some factors on the topology structure that influence on the flow of information between the particles. Kennedy and Mendes have shown that when the average distance between nodes are too short, there is a tendency for the population to move quickly toward the best solution found in earlier iterations [10]. For simple unimodal problems, it usually implies in a faster convergence to the global optimum. However, this fast convergence might be premature in a local optimum, specially in multimodal problems [3]. In this case, communication topologies with lower number of connections may reach better results [3].

The Global topology, as known as $\mathbf{G}_{best}$, was the first topology proposed for the PSO [9]. In the $\mathbf{G}_{best}$, all the particles of the swarm are neighbors, as shown in Figure 1(a). Thus, the social memory of the particles is shared by the entire swarm.

On the other hand, in local topologies, each particle only shares information with a subset of the swarm. Therefore, the social memory is not the same for the whole swarm. The most used local topology is the Ring topology [3], where each particle has only two neighbors, as depicted in the Figure 1(b). This structure helps to avoid a premature attraction of all particles to a single spot of the search space, once the information is spread slowly, but with the caveat of a slow convergence [3].

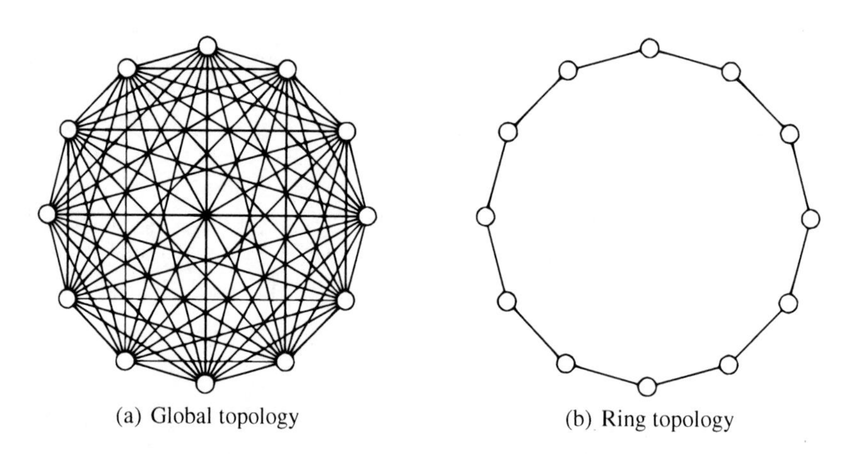

(a) Global topology (b) Ring topology

Fig. 1 Most-used particle swarm optimization communication strategies are based on global and local topologies

These two topologies lead to extreme behaviors in the swarm, therefore many efforts have been made to propose approaches that present fast convergence while avoiding local minima [3, 10]. Some topologies that can self-adapt dynamically were proposed recently [13, 12], including the approach proposed by Oliveira-Júnior *et al.* [12], which is based on the preferential attachment mechanism present on the Barabási-Albert model [1]. In this case, each particle tries to find better particles to create connections with. We will call this topology as dynamic topology.

The Dynamic topology is based on the state of the swarm and it tries to change the information flow only when it is necessary. Because of this, a new attribute, called P_k*failures*, was included in each particle to determine the number of times a particle k does not improve its fitness. If it reaches a pre-determined threshold, then the particle is considered stagnated. If the particle k does not improve its position in the current iteration, P_k*failures* is incremented, otherwise P_k*failures* is set to zero.

The Dynamic topology is initialized as a ring topology. At each iteration of the PSO, all particles update their P_k*failures* and when a preset threshold of failures is reached, the particle searches for better particles to follow, and to stop following as well. This selection of new neighbors is based on a roulette wheel with a fitness-based rank. More details about the dynamic topology can be found in [12].

2.2 Network Science Metrics

The networks discussed in this paper are modeled as graphs. A graph G consists of a pair $[V(G),E(G)]$ where set $V(G)$ set of vertices and $E(G)$ is a set of edges. Any undirected unweighted graph G can be represented by its adjacency matrix $A(G)$, in which the non-diagonal entries (i,j) are equal to "1" if the nodes i and j are adjacent (connected), or "0" otherwise. In A, the entries (i,i) are always equal to "0", because a node cannot be connected to itself. A diagonal matrix, which contains information about the degrees of the nodes, is named Node Degree matrix, $D(G)$. The diagonal entries (i,i) are equal to the degree of the nodes D_i. The Laplacian matrix of a graph G, represented as $L(G)$, is $L(G) = D(G) - A(G)$.

Many properties of a graph can be inferred by using the Laplacian matrix. The eigenvalues $\lambda_1 \leq \lambda_2 \leq ... \leq \lambda_n$ of L are important because they relate to many of these graph properties. Given that L is symmetric, all eigenvalues are real and non-negative. One of the first properties that arises from the Laplacian eigenvalues is the number of components in a graph. The number of zero eigenvalues corresponds exactly to the number of independent sub-graphs in the graph.

The second-smallest eigenvalue λ_2 of L is called the algebraic connectivity (or Fiedler value) of G. The magnitude of this value shows how well connected the graph is. Moreover, the value is greater than 0, if and only if, G is a connected graph.

The Adjacency matrix can also be used to provide information about the structure of the graph. The spectrum of a graph can be defined as the set of eigenvalues of its Adjacency matrix. Assuming this, the spectral density of a graph can be defined as the density of these eigenvalues and can be stated as a probability density function shown in Equation 5.

$$\rho(\lambda) = \frac{1}{N} \sum_{j=1}^{N} \delta(\lambda - \lambda_j). \tag{5}$$

Farkas *et al.* [7] showed that topological features of some kinds of graphs (uncorrelated random networks, the small-world networks, and the scale-free networks) can be identified by its graph spectral density. They also have presented some practical tools for the identification of basic types of random graphs and for classification of real-world graphs [7]. These tools are based on the extremal eigenvalues of the

Adjacency matrix. The extremal eigenvalues contain useful information of the structure of the graph. The principal eigenvalue is detached from the rest of the spectrum depending on the periodicity of the graph structure. Thus, they proposed a quantity named R, defined as:

$$R = \frac{\lambda_1 - \lambda_2}{\lambda_2 - \lambda_N}, \tag{6}$$

that measures the distance of the first eigenvalue from the main part of $\rho(\lambda)$.

The R-value can be used in order to distinguish between some graph-structure features: (i) periodical or almost periodical (Small world); (ii) uncorrelated and non-periodical; and (iii) strongly correlated non-periodical (Scale Free).

3 Simulation Setup and Results

3.1 PSO Setup

In order to assess the proposed methodology, we used a multimodal benchmark function, named *F6 function*, that was proposed in [14] as a large scale optimization problem. The *F6 function* is a single-group shifted and m-rotated Ackley's function. In all experiments, we used $1,000$ dimensions and m equal to 50. We used 200 particles in all simulations. For each simulation trial, we used $300,000$ fitness function evaluations. We performed simulations for the global, local and dynamic topologies. In the case of the dynamic topology, the threshold of failures for the particles was set to $P_k failures = 50$. The particles were updated according to the Equation 4 with $c_1 = 2.05$ and $c_2 = 2.05$ as indicated in [4].

3.2 The Swarm Influence Graph

Although the communication scheme defines which particles can communicate with one another, this does not mean that a particle actually obtains useful information from all the connected particles. Here, we define useful information as the instantaneous use of the $\mathbf{N}_{best_i}(t)$ by the neighbor particle. Since we aim to assess the flow of useful information, we propose here the concept of influence graph. The influence graph consists of a set of trees and each tree represents an information flow through the swarm. Therefore, the number of trees and their structures have a great significance on the swarm behavior. As an example, lots of trees means that there are many different and independent information flows.

The influence graph is defined at each iteration considering only the active links, *i.e.* the ones in which a useful information was provided. One can observe that the influence graph I_i (where i represents an iteration of the execution) is a directed graph by definition. However, in this paper, the edge direction is removed from the graph in order to simplify the analysis and to make use of some available metrics.

Figure 2 depicts examples of influence graphs (in bold) over the three different swarm topologies studied in this paper. One can observe that there are some

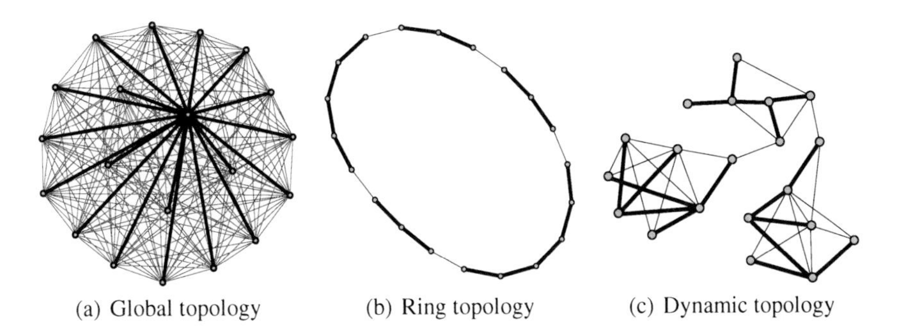

Fig. 2 Examples of influence graphs over the topology for the three communication topologies

sub-graphs for the Ring and Dynamic topologies, whereas the global topology presents a focal point in the influence graph.

3.3 Number of Zero Eigenvalues of the Laplacian Matrix

The number of independent components in the influence graph means the number of information flows within the swarm at the current iteration. As mentioned in Section 2.2, the number of zero eigenvalues of the Laplacian matrix corresponds to the number of sub-graphs in the graph. Therefore, the number of eigenvalues of the Laplacian matrix with value equal to zero indicates the connectivity of the Influence matrix. Figure 3 shows the behavior of this quantity as a function of the number of iterations for the three considered topologies.

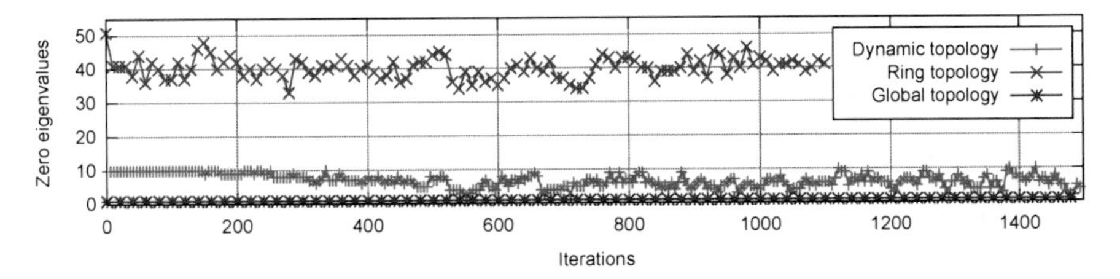

Fig. 3 Number of zero eigenvalues in the Influence matrix for Dynamic, Ring and Global Topologies

The Influence graph in the Global topology is an one-component star-like graph and keeps its structure during the whole execution of the algorithm; the number of zero eigenvalues is constant and equal to one. Although the value is equal to one in this case, it does not mean that the influence graph is the same along the entire process. The best particle of the swarm can change along the iterations which causes the center of the star-like topology to change.

Although the Ring topology is static, the Influence matrix can present different sub-graphs. One can observe that the number of information flows varies through the iterations, but presents a high average value of 40 along the entire process.

In the Dynamic topology, the algorithm starts with 10 information flows and diminishes in average to 5 information flows along the algorithm execution. One must observe that the Dynamic topology presents a balanced behavior between the two static approaches (Global and Ring).

3.4 *R-value*

As described in Section 2.2, the R-value represents a relation between important eigenvalues. Figure 4 presents the behavior of R-value of the Influence graph as a function of the number of iterations for the three considered topologies. A low R-value means that λ_1 is not detached from the rest of the spectrum and it can be seen as a consequence of a periodical structure [7].

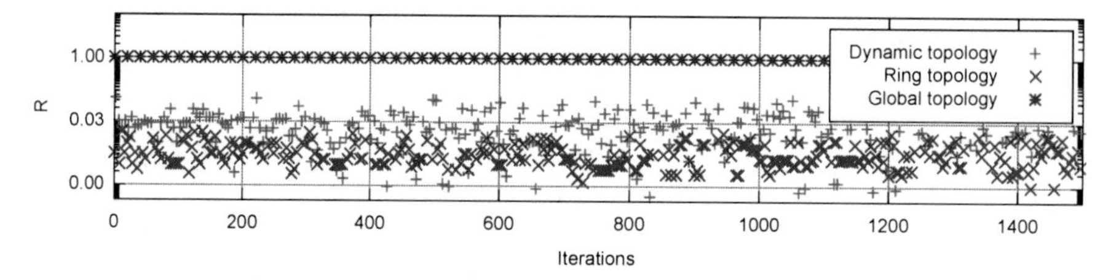

Fig. 4 R-value of the Influence matrix for Dynamic, Ring and Global Topologies

Because of the star-like behavior of influence graph for the Global topology, its R-value is constant and presents the maximum possible value (1), since the extremal eigenvalues λ_1 and λ_N are opposites.

Both the Ring and Dynamic topologies present small R-values, as they produce influence graphs that features quite periodical structures. Again, the Dynamic topology presented a balanced behavior.

3.5 *The Density Spectrum*

The spectral density has the capacity to represent the frequency of the eigenvalues. Therefore, it is interesting to evaluate the characteristics of the topologies as a function of the number of iterations.

As the Ring and Global topologies present a well known behavior in terms of connectivity and convergence, we first evaluated the evolution of the spectral density of the influence graph as a function of the number of iterations for these two topologies. The results for the iterations 100, 400, 800 and 1200 are depicted in Figure 5. One can observe that the Ring topology presents a bi-modal shape, while the

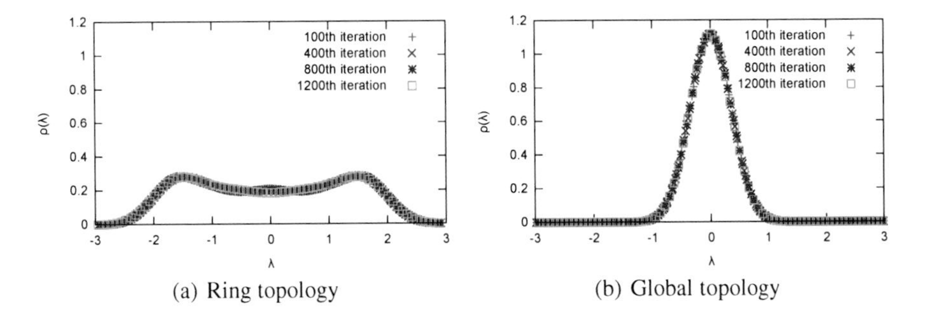

(a) Ring topology (b) Global topology

Fig. 5 Density spectrum of the swarm influence graph for the statical topologies

Global topology presents a perfect uni-modal shape. Besides, the shapes are very well defined and do not vary along the iterations.

After that, we evaluated the Dynamic topology. Figure 6 shows two different trials of the PSO with the dynamic topology (Run #1 and Run #2). As can be observed, the results for these two independent runs were quite different. In Run #1, the algorithm converged, while in the Run #2 the algorithm got stuck in a local minima. Figure 7 shows the spectral density of the evolution of the influence graph for these two runs for the iterations 100, 400, 800 and 1200. In Run #2, the swarm probably got stuck in a local minima because the topology presented a Global behavior at the beginning of the execution, *i.e.* the shape of the spectral density is perfectly uni-modal around iteration number 100.

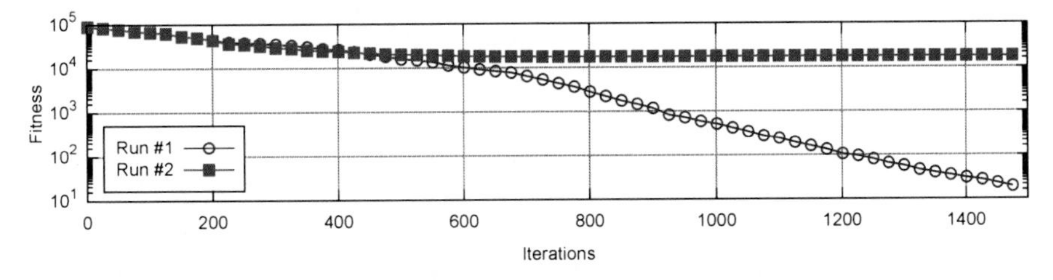

Fig. 6 Two independent runs of the PSO algorithm with the Dynamic topology

3.6 P_k*failures* Threshold Impact

In order to show that we can use our methodology to carry out deeper analyses, we studied the influence of P_k*failures*, which plays a important role in the performance of the Dynamic topology. In general, if P_k*failures* has a low value, the particles will easily try to reconnect. Otherwise, particles will tend to maintain the current topology. One must observe that this value also has an impact on the spectral density of the influence graph.

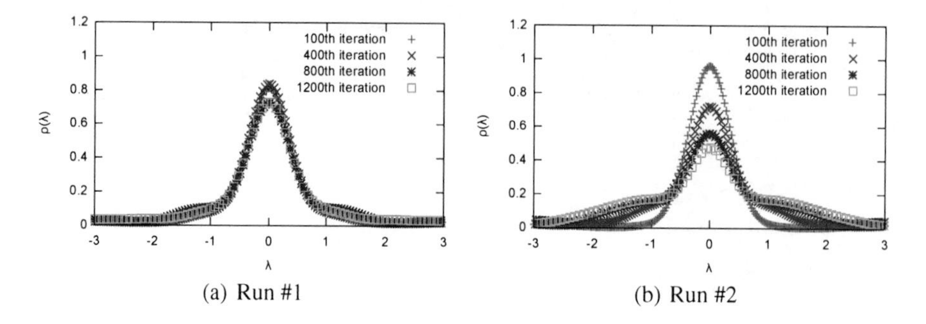

(a) Run #1 (b) Run #2

Fig. 7 Density spectrum of the swarm influence graph in different algorithm executions

Figure 8(a) shows the fitness evolution through iterations for $P_k failures$ equal to $1, 5$ and 50. Figure 8 depicts the density spectra for the three cases. One can observe that in all cases the spectral densities present a combination of the uni-modal and bi-modal curves. The fitness obtained for $P_k failures = 1$ was not satisfactory because, in this case, the topology changes a lot and it diminishes the convergence capability.

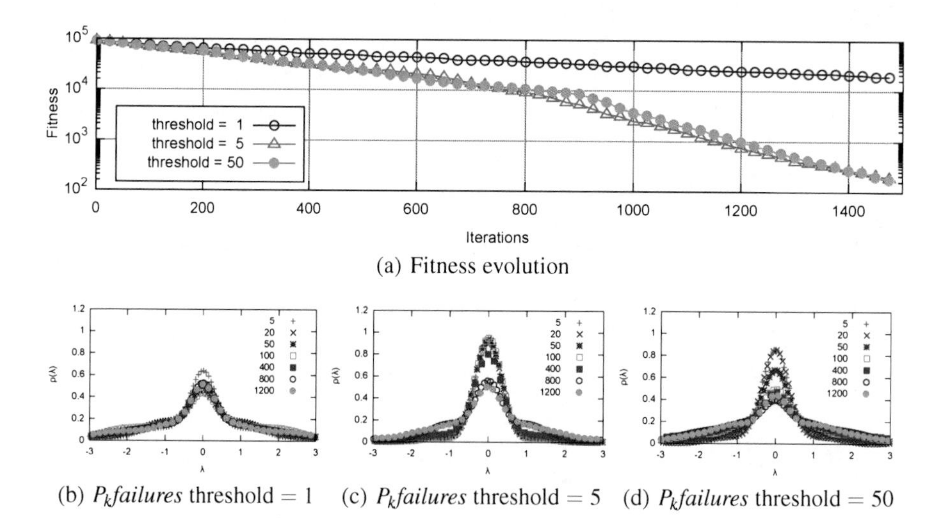

(a) Fitness evolution

(b) $P_k failures$ threshold $= 1$ (c) $P_k failures$ threshold $= 5$ (d) $P_k failures$ threshold $= 50$

Fig. 8 Fitness evolution and density spectrum of the swarm influence graph with different $P_k failures$ thresholds

4 Conclusions and Future Works

We proposed in this paper a set of tools based on some Network Science metrics to assess the information flow on the Particle Swarm Optimization algorithms. We observed that it is necessary to assess the influence graph, instead of the topology itself in order to evaluate only the flow of useful information. The tools link the structure features of the influence graph to algebraic quantities, that may be used for further analysis in order to understand the swarm behavior.

We have shown that the swarm with a Dynamic topology has a behavior that is between the two most used static approaches. Moreover, the simulation results indicate that the stagnation can be foreseen by analyzing the density spectrum along the iterations.

As future works, we intend to use these tools to design high performance dynamic topologies for PSOs by assessing the information flow within the swarm. We also aim to develop variations of the swarm influence graph to recognize different aspects of the swarm communication.

References

1. Albert, R., Barabasi, A.L.: Statistical mechanics of complex networks. Reviews of Modern Physics 74, 47 (2002), doi:doi:10.1103/RevModPhys.74.47
2. Bastos-Filho, C.J.A., Lima-Neto, F.B., Lins, A.J.C.C., Nascimento, A.I.S., Lima, M.P.: A novel search algorithm based on fish school behavior. In: 2008 IEEE International Conference on Systems, Man and Cybernetics, pp. 2646–2651 (2008)
3. Bratton, D., Kennedy, J.: Defining a standard for particle swarm optimization. In: Swarm Intelligence Symposium, SIS 2007, pp. 120–127. IEEE (2007), doi:10.1109/SIS.2007.368035
4. Clerc, M., Kennedy, J.: The particle swarm - explosion, stability, and convergence in a multidimensional complex space. IEEE Transactions on Evolutionary Computation 6(1), 58–73 (2002), doi:10.1109/4235.985692
5. Dorigo, M., Caro, G.: Ant colony optimization: A new meta-heuristic. In: Proceedings of the Congress on Evolutionary Computation, pp. 1470–1477. IEEE Press (1999)
6. Engelbrecht, A.P.: Computational Intelligence: An Introduction. Wiley Publishing (2007)
7. Farkas, I.J., Derényi, I., Barabási, A.L., Vicsek, T.: Spectra of "real-world" graphs: beyond the semicircle law. Phys Rev E Stat Nonlin Soft Matter Phys 64(2 Pt 2) (2001), `http://view.ncbi.nlm.nih.gov/pubmed/11497741`
8. Karaboga, D.: An idea based on honey bee swarm for numerical optimization. Tech. rep., Erciyes University, Engineering Faculty, Computer Engineering Department (2005)
9. Kennedy, J., Eberhart, R.: Particle swarm optimization, vol. 4, pp. 1942–1948 (1995), `http://dx.doi.org/10.1109/ICNN.1995.488968`, doi:10.1109/ICNN.1995.488968
10. Kennedy, J., Mendes, R.: Population structure and particle swarm performance. In: Proceedings of the 2002 Congress on Evolutionary Computation, CEC 2002, vol. 2, pp. 1671–1676 (2002), doi:10.1109/CEC.2002.1004493
11. Lewis, T.G.: Network Science: Theory and Applications. Wiley Publishing (2009)
12. Oliveira-Júnior, M.A.C., Bastos-Filho, C.J.A., Menezes, R.: Using Network Science to Define a Dynamic Communication Topology for Particle Swarm Optimizers. In: Menezes, R., Evsukoff, A., González, M.C. (eds.) Complex Networks. SCI, vol. 424, pp. 39–47. Springer, Heidelberg (2013), doi:10.1007/978-3-642-30287-9_5

13. Suganthan, P.: Particle swarm optimiser with neighbourhood operator. In: Proceedings of the 1999 Congress on Evolutionary Computation, CEC 1999, vol. 3, xxxvii+2348 (1999), doi:10.1109/CEC.1999.785514
14. Tang, K., Li, X., Suganthan, P.N., Yang, Z., Weise, T.: Benchmark Functions for the CEC'2010 Special Session and Competition on Large-Scale Global Optimization. Tech. rep., University of Science and Technology of China (USTC), School of Computer Science and Technology, Nature Inspired Computation and Applications Laboratory (NICAL): China
15. Watts, D.J., Strogatz, S.H.: Collective dynamics of 'small-world' networks. Nature 393(6684), 440–442 (1998), http://dx.doi.org/10.1038/30918, doi:10.1038/30918

Robustness of Network Controllability under Edge Removal

Justin Ruths and Derek Ruths

Abstract. We introduce a quantitative measure of robustness of network controllability. Given a set of control nodes which drive the network, we investigate the effect of edge removal on the number of controllable nodes. We find that the mean degree of the network is a major factor in determining the robustness of random networks. Nonetheless, a comparison between real and random networks indicates a statistically significant difference which points to additional factors that influence the robustness of control of real complex networks.

1 Introduction

Mounting evidence confirms that network-oriented studies can uniquely characterize aspects of complex dynamics in wide ranging areas of biology, engineering, and social science: e.g., interaction of proteins in cellular pathways, diffusion of information within human communities, and optimal routing strategies on the internet [9, 4, 5]. However, as researchers seek techniques for engineering or influencing such complex systems, they are confronted by questions of controllability.

Real-world networks bring with them their own unique set of attributes and requirements not found in typical control systems. For example, in real complex networks (e.g., cellular biochemical pathways, social networks, and sprawling internet physical connectivity) external controls are not predefined and mechanisms for controls are not known, a priori. Furthermore, the structure of such systems is unlikely to be static over time and space: cascades of fundamental biochemical interactions

Justin Ruths
Singapore University of Technology and Design, 20 Dover Drive, Singapore
e-mail: justinruths@sutd.edu.sg

Derek Ruths
McGill University, 3480 University Street, Montreal, Canada
e-mail: derek.ruths@mcgill.ca

G. Ghoshal et al. (Eds.): *Complex Networks IV*, SCI 476, pp. 185–193.
DOI: 10.1007/978-3-642-36844-8_18

differ among individuals, friendships may be made or lost, and internet connectivity changes as individual servers come online or fall offline. Thus, a schema for controlling such systems over long periods of time must be robust to these frequent and unpredictable structural changes. Because the addition and removal of edges and nodes in a network has no clear analog to modifications to a standard control system, this problem of robust network control presents a new challenge both to the network and control communities.

In this paper, we first define a quantitative measure of the robustness of a given control scheme for a given network. Second, we develop the necessary computational techniques to use our measure and network sampling to assess the statistical significance of robustness observed in a real-world network. This second result allows us to consider for the first time whether the controllability of real networks is more or less robust than might be expected at random.

2 Background

We consider a linear dynamic model of the directed network graph $G(A)$, which is composed of n nodes and L directed edges between nodes in the network. Systematically analyzing linear models is a key step in generalizing to broader classes of models, such as nonlinear dynamics. In addition, there are a large number of network phenomena that exhibit a good fit to a linear time-invariant model of the form [8, 13, 10], $\dot{x}(t) = Ax(t)$, where the state $x(t) \in \mathbb{R}^n$ is the value of all of the nodes at time t and $A \in \mathbb{R}^{n\times n}$ is the transpose of the adjacency matrix of the network, such that the value $A_{i,j}$ is the weight of a directed edge from node x_j to node x_i and zero if there is no such edge. In what follows, we study the controlled network $G(A,B)$, corresponding to adding m control nodes yielding the form, $\dot{x}(t) = Ax(t) + Bu(t)$ where the control $u(t) \in \mathbb{R}^m$ and $B \in \mathbb{R}^{n\times m}$ models the effect of the controls on the network. Determining the minimal B to make the graph $G(A,B)$ completely controllable has been a topic of particular interest [11, 2, 3]. A networked system is controllable if the controls are able to guide the system state, the value of the nodes, from an initial configuration $x(t_0) = x_0 \in \mathbb{R}^n$ to a final configuration $x(t_1) = x_1 \in \mathbb{R}^n$. The minimum number of control nodes which makes the $G(A,B)$ completely controllable (m_c) has been shown to correlate with the degree distribution, which is a measure of edges leaving nodes of the network [8].

Due to space constraints, we are unable to review the terms that arise in the following sections. These terms are well defined by a number of sources, however, they are likely new to many, even well-versed, network scientists [7, 11, 2, 3]. In particular, we direct you to review the concept of structural controllability and its relation to generic rank of the system $[A\ B]$; cacti structure of a network formed by paths and cycles of connected nodes; and maximum matching algorithm to find the largest set of nodes that can be uniquely paired amongst themselves using edges present in the network.

3 Methods

We discuss two novel measures of the robustness of a control scheme given by B to changes in the network structure through edge removal. The first is a measure that assigns a value to a network/control scheme pair based on how reachability responds to the removal of edges from the network. The second is a computational approach to estimating the statistical significance of a network/control scheme pair's robustness. This answers the question: how much more robust is the network/control pair than might be expected at random?

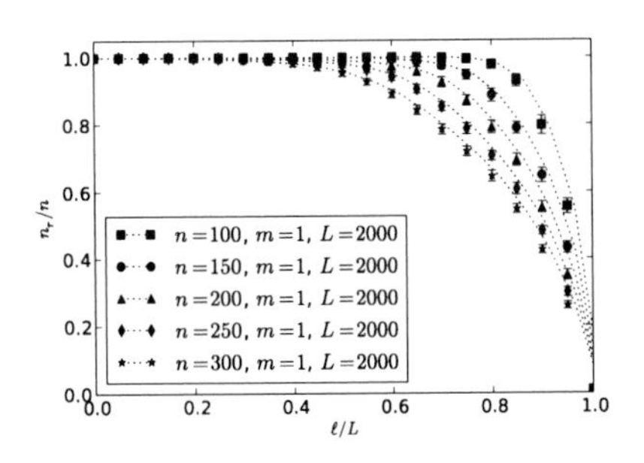

Fig. 1 The robustness profiles of Erdos-Renyi networks with the same number of original edges. The number of controllable nodes n_r of five ER networks are shown as a function of the number of edges, ℓ, removed. The displayed values are the averages of 50 separate percolation sequences, with corresponding error bars. The robustness measure, R, is given as the normalized area under this curve.

3.1 Measuring Robustness

Under fully structural controllable conditions, all nodes in the network are reachable by the specified control scheme. However, as edges are removed from the network (hereafter, *percolated*) some nodes will become unreachable. For a given sequence of edge percolations, the number of unreachable nodes will depend on the location of the controls in the network and the local and global structure of the network itself. Conceptually, given two different control schemes and a sequence of edge percolations, the more robust control scheme is the one that maintains a greater number of reachable nodes in the percolated network (in which the edges are missing).

In order to derive a quantitative measure of robustness, first consider that we can incrementally remove edges from a network with a specified control scheme and compute the number of nodes in the network that are still controllable. Such an iterative calculation will yield curves like those shown in Figure 1, which we call *robustness profiles*. In such a plot, the x-axis is the fraction of edges that have been percolated and the y-axis is the fraction of nodes that are reachable in the percolated network. The value of point ℓ/L corresponds to the fraction of nodes that are reachable after removing ℓ/L percent of the edges present in the original network. Curves will be monotonically decreasing since removing an edge cannot bring a new node under control.

Every combination of network (G), control scheme (B), and edge percolation sequence (ξ) has its own robustness profile, which we can summarize by integrating to find the area under the curve. We call this value the *integrated percolation (IP) robustness*,

$$R_{G,B,\xi} = \frac{1}{nL} \sum_{\ell=0}^{L} n_r(G - \xi[1:\ell], B) \tag{1}$$

where G is the network, n is the number of nodes in the network, L is the number of edges, B is the control scheme, $n_r(G', B')$ is the number of reachable nodes in the graph G' with controls B', and $\xi[1:\ell]$ is the first ℓ edges in the edge percolation sequence. $G - \xi[1:\ell]$ is graph G with the first ℓ edges in ξ removed.

Because the robustness profile always falls within a unit box (i.e., both axes of the plot are normalized to 1), the intuitive interpretation of $R_{G,B,\xi}$ is as the percent of the unit box that falls under the curve. This has the desired effect of returning larger values for more robust control schemes: if we fix the network and edge percolation sequence, a control scheme that retains more reachable nodes will have a curve with more of the area of the unit box under it.

Given the ability to compute the IP robustness for arbitrary edge percolation sequences, we can obtain an estimate of the robustness of a network-control scheme pair (hereafter, a *configuration*) by averaging $R_{G,B,\xi}$ over a large number of randomly generated edge percolation sequences, $\Xi = \{\xi_1, \xi_2, \ldots\}$,

$$R_{G,B} = \frac{1}{|\Xi|} \sum_{\xi \in \Xi} R_{G,B,\xi}. \tag{2}$$

We call this measure the *mean integrated percolation (MIP) robustness*. Analysis on a wide spectrum of networks have shown that, in practice (see Figures), the value of $R_{N,B,\xi}$ is quite consistent across different percolation sequences (i.e., the standard deviation for the statistic is diminishingly small). This consistency lends $R_{G,B}$ to being a reliable measure of robustness of a network-control scheme pair to arbitrary edge percolation processes. This will be further explored and validated in Section 4.

Hereafter, mentions of robustness refer to MIP robustness. Standard deviation will be shown in plots to indicate the variability in the estimate.

3.2 *Assessing Real Network Robustness*

In the preceding subsections, we were interested in simply being able to compute and compare the robustness of different network-control configurations. Such techniques are not sufficient, however, to quantify the extent to which real networks are designed for robust control. The notion of "designed robustness", whether through engineering or evolution, suggests that such networks would be expected to have more robustness than might be observed by chance. In this subsection we formalize this notion and provide a method for estimating the probability of observing a given network's MIP robustness by chance.

A standard practice in network science for establishing the statistical significance of a given network feature is to compare the feature in the network of interest to the same feature in a set of synthetic networks drawn from a null model (e.g., [6, 9]). In this instance, we will compare the robustness of the network of interest, G_0, to the robustness of a set of randomized networks, $\mathbf{Z} = \{G_1, G_2, ...\}$. In order to facilitate comparison, the random networks must match the number of nodes and edges in the original network ($n_i = n_0$ and $L_i = L_0$ for $i \geq 1$). Furthermore, we preserve the degree distribution of the original network in the random networks: this is done because, as we will see, mean degree is strongly associated with narrow ranges of robustness. Thus, the degree distribution of the network alone can explain much of its robustness. By holding degree distribution constant, our results estimate the extent to which more higher-order degree features and local motifs present within the real networks lend it to robust control.

To perform an actual comparison, a large number of random networks satisfying the above constraints are generated. The average MIP robustness is computed for the null model for a range of control scheme sizes: $R_{\mathbf{Z}}(m) = \frac{1}{|\mathbf{Z}|} \sum_{G \in \mathbf{Z}} R_{G,B_m}$. The MIP robustness for the original network is also computed for a range of control scheme sizes: $R_0(m) = R_{G_0,B_m}$. For a given value of m, $R_{\mathbf{Z}}(m)$ and $R_0(m)$ can be directly compared. Furthermore, since both measures are means, the overlap in their standard deviations can be used to assess the statistical significance of the original network's MIP robustness score.

4 Results and Discussion

In this section, we apply our definitions and methods to understanding two specific questions. First, we quantify the robustness of a class of synthetic networks and explore the connection between degree distribution and MIP robustness. Second, we determine the extent to which real networks are more robust than might be expected by chance, which is a first step towards understanding whether robust control is a feature designed into some complex systems.

4.1 Random Networks

Random network models have the capacity to generate large numbers of networks which share certain properties, but are random in every other way. For this reason, such network models have been fruitfully used to study the impact of specific network properties on phenomena of interest. Here we employ them for similar purposes: to understand the extent to which network properties influence MIP robustness. There are many different random network models; in the present paper we focus on Erdos-Renyi (ER) networks which are generated by cycling through all pairs of nodes and establishing an edge between these nodes with probability p [1]. The number of edges in an ER network is, on average, $L = n^2 p$.

Figure 1 depicts the effect of edge removal on the number of controllable nodes. The generated ER networks have the same number of edges ($L = 2000$) with varying number of total nodes ($n \in \{100, 150, 200, 250, 300\}$). In addition the number of controls is kept constant, $m = 1$, which is crucial for a fair comparison. The robustness profiles in Figure 1 are the result of averaging 50 percolation sequences for each network, as described by Equation 2. If we start with a control scheme (B) that makes $G(A,B)$ fully controllable, the robustness profile will start with $n_r(0) = n$ and end at $n_r(L) = m$. Although the curves in Figure 1 are derived for a single realization of each ER network, we have observed that any ER network with the same n and L parameters will result in effectively indistinguishable robustness profiles, which we omit here for brevity.

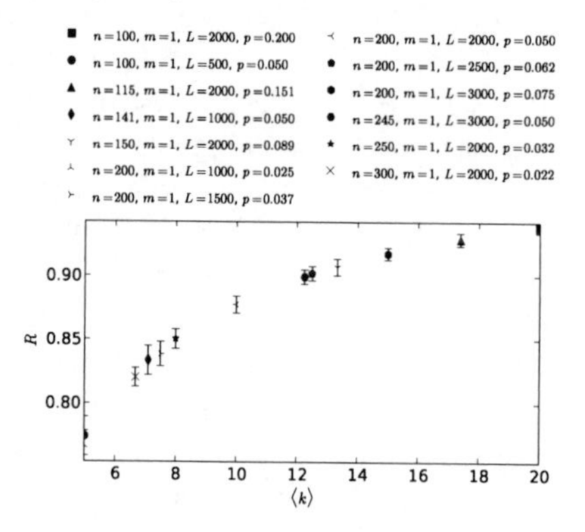

Fig. 2 The MIP robustness of Erdos-Renyi networks. We compute the robustness profiles and the corresponding MIP robustness measures, $R_{G,B}$, for a variety of ER networks. We plot the robustness measure versus the mean degree $\langle k \rangle = L/n$.

Figure 1 shows that changing the number of nodes changes the curvature of the robustness profile, with lower numbers of nodes yielding a higher MIP robustness measure, as given by Equations 1 and 2. This result agrees with our intuition based on cacti-based control scheme construction: if the network is more dense, there are more edges for each node. In this case, it is more likely that m stems (node-disjoint paths) will be able to cover a larger part of the network. Moreover, more edges per node increase the chance that an alternative path exists around a location where an edge was removed.

With this insight we now aim to characterize the property (or properties) that help to determine the robustness of a controlled network. We now broaden the parameters used to generate the ER networks. In addition to those shown in Figure 1 we include several ER networks with the same number of nodes ($n = 200$) and also several with the same density ($p = 0.05$). Again we compute the robustness profiles and the corresponding MIP robustness measures for 50 percolation sequences. These were plotted against several choices of network parameters, such as n, L, and p. However, none had as strong a correlation as the mean degree $\langle k \rangle$, which is shown in Figure 2. This strong correlation supports the idea that there may be a causal relationship between degree-based network properties and the network's MIP robustness. More generally, the presence of a strong correlation with mean degree confirms that aspects of network structure can, indeed, influence the robustness of controllability, suggesting that other features besides mean degree may also confer some amount of robustness.

4.2 *Real Networks*

Although random networks can reveal certain underlying concepts of control robustness, our ultimate goal is to quantify, characterize, and engineer robustness in real-world systems. Fundamental to this investigation is determining whether real networks are more or less robust than might be expected by chance. For example, do certain systems have robust controllability as a design criterion whereas others sacrifice this in favor of other attributes such as performance, adaptability, and efficiency? We study the difference in MIP robustness between several real networks and their cooresponding degree-preserving random networks, including a food web, email correspondence (the East Anglia email dataset), and protein networks [14, 12]. We chose these networks because the concept of controllability is directly relevant to each: controls can disrupt or correct disruptions within the food chain of an ecosystem, influence the spread of information among a population, and be used to alter the trajectory of biochemical systems.

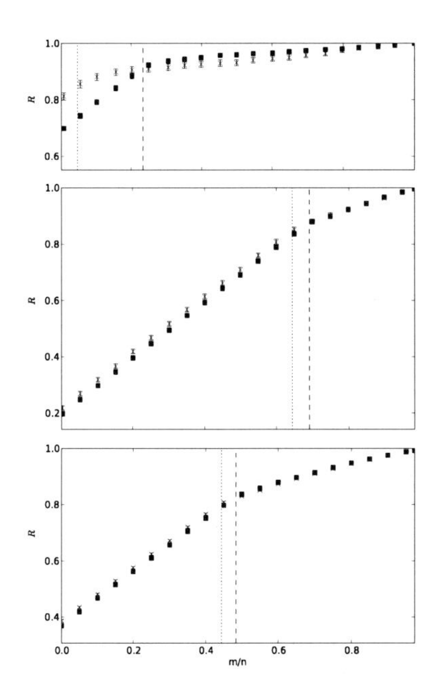

Fig. 3 The MIP robustness of real networks. In order from top to bottom, we compute robustness profiles and the corresponding MIP robustness measures (■) for a food web, email correspondence, and yeast protein networks. We compare these results with degree-preserving shuffled networks (×). The dashed and dotted vertical lines correspond to the number of controls, m_c, needed to fully control the real and shuffled networks, respectively. A t-test confidence level beyond 0.05 confirms the definitive switch from shuffled to real networks being more controllable before and after the dashed line, respectively. For the food web and yeast plots this confidence is beyond 0.0005.

The comparison of robustness between two arbitrarily sized and controlled networks is challenging. In order to make such a comparison possible between a real and a random network, we maintained the degree distribution of the network as well as its number of nodes and edges. We further consider the MIP robustness differences for a range of numbers of controls (m). In Figure 3 we show the MIP robustness for both real and shuffled networks, constraining the control set to a size m, where m ranges between 1 to n. The dashed and dotted vertical lines correspond to the number of controls, needed to fully control the real (m_c^{real}) and shuffled (m_c^{shuff}) networks, respectively. For all $m < m_c^{\text{real}}$, we observe that the shuffled networks (denoted by $\times$) are consistently more robust than the original real network (denoted by ■). However, after m_c^{real}, the relative robustness is reversed and the real networks are more robust. In a statistical t-test of the MIP robustness values, this switch is confirmed with a confidence beyond 0.05 for all tested networks (note that as $m/n \to 1$, the robustness difference disappears since both networks approach fully controllable even under percolation). This confidence is upheld prior to the dashed line and several nodes past it until the measures begin to coincide at the far right. In the upper and lower subplots, corresponding to the food web and yeast protein networks, the trends have a statistical significance beyond 0.0005.

These observations provide an intriguing starting point for deeper investigation into the basis and origins of robust control in real networks. The fact that the switching behavior is both significant and conserved across networks drawn from such different real-world systems suggests that certain kinds of control robustness may be preferred by quite diverse natural systems. For example, the fact that robustness is enriched only for $m \geq m_c^{\text{real}}$, may imply that real systems are trading off robustness for other desirable attributes such as efficiency or adaptability. The fact that these real networks are more robust than networks drawn from degree-preserving random models suggests that more subtle and, potentially, local structures are involved in implementing robust control.

5 Conclusion and Future Work

Where complex systems are useful or relevant to human pursuits, there will be an interest in the ability control and influence them. Because many natural systems experience frequent and unpredictable structural changes, we seek controls that are robust to changes in the structure of the controlled system.

In this paper, we have approached the issue of characterizing and comparing the robustness of networks. We have outlined a methodology for investigating the robustness of networks under edge removal, which augments the current methods in structured systems and graph theory literature.

There are a number of promising directions for future work in this area. In order to compare two real-world networks, we must develop a formal way of comparing robustness between control configurations of different sizes. Additionally, in order to understand the ways in which control robustness can be constructed, a careful assessment is needed of the contributions that different network properties (e.g., clustering, centrality, motifs) make to the robust controllability of a network.

References

1. Erdos, P., Renyi, A.: On random graphs. Publicationes Mathematicae 6, 290–297 (1959)
2. Glober, K., Silverman, L.M.: Characterization of structural controllability. IEEE Transactions on Automatic Control AC-21(4), 534–537 (1976)
3. Hosoe, S., Matsumoto, K.: On the irreducibility condition in the structural controllability theorem. IEEE Transactions on Automatic Control AC-24(6), 963–966 (1979)
4. Kitano, H.: Systems biology: A brief overview. Science 295(5560), 1662–1664 (2002)
5. Lazer, D., Pentland, A., Adamic, L., Aral, S.: Life in the network: the coming age of computational social science. Science 323(5915), 721–723 (2009)

6. Li, F., Long, T., Lu, Y., Ouyang, Q., Tang, C.: The yeast cell-cycle network is robustly designed. Proceedings of the National Academy of Sciences of USA 101(14), 4781–4786
7. Lin, C.-T.: Structural Controllability. IEEE Transactions on Automatic Control AC-19(3), 201–208 (1974)
8. Liu, Y.-Y., Slotine, J.-J., Barabasi, A.-L.: Controllability of complex networks. Nature 473, 167–173 (2011)
9. Milo, R., Shen-Orr, S., Itzkovitz, S., Kashtan, N., Chklovskii, D., Alon, U.: Network motifs: Simple building blocks of complex networks. Science 298(5594), 824–827 (2002)
10. Rahmani, A.R., Ji, M., Mesbahi, M., Egerstedt, M.B.: Controllability of multi-agent systems from a graph-theoretic perspective. SIAM Journal on Control and Optimization 48(1), 162–186 (2009)
11. Shields, R.W., Pearson, J.B.: Structural controllability of multi-input linear systems. IEEE Transactions on Automatic Control AC-21(2), 203–212 (1976)
12. Sun, S., Ling, L., Zhang, N., Li, G., Chen, R.: Topological structure analysis of the protein-protein interaction network in budding yeast. Nucleic Acids Research 31(9), 2443–2450 (2003)
13. Tanner, H.G.: On the controllability of nearest neighbor interconnections. In: CDC 43rd IEEE Conference, vol. 3, pp. 2467–2472 (2004)
14. Ulanowicz, R.E., Bondavalli, C., Egnotovich, M.: Network analysis of trophic dynamics in south florida ecosystem, fy 97: The florida bay ecosystem. Technical Report UMCES-CBL 98-123, Chesapeake Biological Laboratory (1998)

Author Index